Sm05002771

KU-266-942

621·389 28 (Luh)

WITHDRAWN

WITHDRAWN

WITHDRAWN

CARLISLE LIBRARY
ST MARTINS SERVICES LTD.

Register Your Book

at www.phptr.com/ibmregister/

Upon registration, we will send you electronic sample chapters from two of our popular IBM Press books. In addition, you will be automatically entered into a monthly drawing for a free IBM Press book.

Registration also entitles you to:

- Notices and reminders about author appearances, conferences, and online chats with special guests

- Access to supplemental material that may be available

- Advance notice of forthcoming editions

- Related book recommendations

- Information about special contests and promotions throughout the year

- Chapter excerpts and supplements of forthcoming books

Contact us

If you are interested in writing a book or reviewing manuscripts prior to publication, please write to us at:

Editorial Director, IBM Press
c/o Pearson Education
One Lake Street
Upper Saddle River, New Jersey 07458

e-mail: IBMPress@pearsoned.com

Visit us on the Web: www.phptr.com/ibmpress/

RFID Sourcebook

IBM Press

DB2® BOOKS

DB2® Universal Database V8 for Linux, UNIX, and Windows Database Administration Certification Guide, Fifth Edition
Baklarz and Wong

Understanding DB2®
Chong, Liu, Qi, and Snow

Integrated Solutions with DB2®
Cutlip and Medicke

High Availability Guide for DB2®
Eaton and Cialini

DB2® Universal Database V8 Handbook for Windows, UNIX, and Linux
Gunning

DB2® SQL PL, Second Edition
Janmohamed, Liu, Bradstock, Chong, Gao, McArthur, and Yip

DB2® Universal Database for OS/390 V7.1 Application Certification Guide
Lawson

DB2® for z/OS® Version 8 DBA Certification Guide
Lawson

DB2® Universal Database V8 Application Development Certification Guide, Second Edition
Martineau, Sanyal, Gashyna, and Kyprianou

DB2® Universal Database V8.1 Certification Exam 700 Study Guide
Sanders

DB2® Universal Database V8.1 Certification Exam 703 Study Guide
Sanders

DB2® Universal Database V8.1 Certification Exams 701 and 706 Study Guide
Sanders

DB2® Universal Database for OS/390
Sloan and Hernandez

The Official Introduction to DB2® for z/OS®, Second Edition
Sloan

Advanced DBA Certification Guide and Reference for DB2® Universal Database v8 for Linux, UNIX, and Windows
Snow and Phan

DB2® Express
Yip, Cheung, Gartner, Liu, and O'Connell

DB2® SQL Procedure Language for Linux, UNIX and Windows
Yip

DB2® Version 8
Zikopoulos, Baklarz, deRoos, and Melnyk

ON DEMAND COMPUTING BOOKS

Business Intelligence for the Enterprise
Biere

On Demand Computing
Fellenstein

Grid Computing
Joseph and Fellenstein

Autonomic Computing
Murch

RATIONAL

Software Configuration Management Strategies and IBM Rational ClearCase®, Second Edition
Bellagio and Milligan

WEBSPHERE BOOKS

IBM® WebSphere®
Barcia, Hines, Alcott, and Botzum

IBM® WebSphere® Application Server for Distributed Platforms and z/OS®
Black, Everett, Draeger, Miller, Iyer, McGuinnes, Patel, Herescu, Gissel, Betancourt, Casile, Tang, and Beaubien

Enterprise Java™ Programming with IBM® WebSphere®, Second Edition
Brown, Craig, Hester, Pitt, Stinehour, Weitzel, Amsden, Jakab, and Berg

IBM® WebSphere® and Lotus
Lamb, Laskey, and Indurkhya

IBM® WebSphere® System Administration
Williamson, Chan, Cundiff, Lauzon, and Mitchell

Enterprise Messaging Using JMS and IBM® WebSphere®
Yusuf

MORE BOOKS FROM IBM PRESS

Developing Quality Technical Information, Second Edition
Hargis, Carey, Hernandez, Hughes, Longo, Rouiller, and Wilde

Performance Tuning for Linux® Servers
Johnson, Huizenga, and Pulavarty

Building Applications with the Linux Standard Base
Linux Standard Base Team

An Introduction to IMS™
Meltz, Long, Harrington, Hain, and Nicholls

Search Engine Marketing, Inc.
Moran and Hunt

Inescapable Data
Stakutis and Webster

RFID Sourcebook

Sandip Lahiri

IBM Press
Pearson plc

Upper Saddle River, NJ • New York • San Francisco
Toronto • London • Munich • Paris • Madrid
Capetown • Sydney • Tokyo • Singapore • Mexico City

www.phptr.com/ibmpress

The author and publisher have taken care in the preparation of this book, but make no expressed or implied warranty of any kind and assume no responsibility for errors or omissions. No liability is assumed for incidental or consequential damages in connection with or arising out of the use of the information or programs contained herein.

© Copyright 2006 by International Business Machines Corporation

Note to U.S. Government Users: Documentation related to restricted right. Use, duplication, or disclosure is subject to restrictions set forth in GSA ADP Schedule Contract with IBM Corporation.

IBM Press Program Manager: Tara Woodman, Ellice Uffer
IBM Press Consulting Editor: Karen Keeter
Cover Design: IBM Corporation

Published by Pearson plc
Publishing as IBM Press

Library of Congress Catalog Number
2005926406

IBM Press offers excellent discounts on this book when ordered in quantity for bulk purchases or special sales, which may include electronic versions and/or custom covers and content particular to your business, training goals, marketing focus, and branding interests. For more information, please contact:

> U.S. Corporate and Government Sales
> 1-800-382-3419
> corpsales@pearsontechgroup.com.

For sales outside the U.S., please contact:

> International Sales
> international@pearsoned.com.

The following terms are trademarks or registered trademarks of International Business Machines Corporation in the United States, other countries, or both: DB2, Lotus, Tivoli, WebSphere, z/OS, Rational, IBM, the IBM logo, and IBM Press. Java and all Java-based trademarks are trademarks of Sun Microsystems, Inc., in the United States, other countries, or both. Microsoft, Windows, Windows NT, and the Windows logo are trademarks of the Microsoft Corporation in the United States, other countries, or both. Linux is a registered trademark of Linus Torvalds. Intel, Intel Inside (logo), MMX, and Pentium are trademarks of Intel Corporation in the United States, other countries, or both. OSF/1 and UNIX are registered trademarks and The Open Group is a trademark of The Open Group in the United States and other countries. Other company, product, or service names mentioned herein may be trademarks or service marks of their respective owners.

All rights reserved. This publication is protected by copyright, and permission must be obtained from the publisher prior to any prohibited reproduction, storage in a retrieval system, or transmission in any form or by any means, electronic, mechanical, photocopying, recording, or likewise. For information regarding permissions, write to:

> Pearson Education, Inc.
> Rights and Contracts Department
> One Lake Street
> Upper Saddle River, NJ 07458

ISBN 0-13-185137-3

Text printed in the United States on recycled paper at Courier in Westford, Massachusetts.
First printing, September 2005

For Anasua, Sohini, and Aishani

Table of Contents

Preface

Radio frequency identification (RFID) is one of the hottest emerging technologies today. Its use has the potential to affect an extremely wide spectrum of the population (from the technology adopters, to vendors, to integrators, to users). Any technology such as RFID raises a simple but important question: How will it be used? At this point, no simple explanation answers this question. Its use will be determined by a variety of factors, including people's interest in this technology and regulations/standards adopted regarding its use.

For example, regulatory bodies (governmental entities) of different countries might impose restrictions on RFID use specific to their country of authority. Likewise, standards bodies (both domestic and international) might prescribe RFID build specifications to address real-world issues. Businesses might look for ways to use RFID to improve customer service. Privacy rights groups might suggest ways to use RFID technology that do not infringe on individual privacy rights. The list goes on.

Currently, the agendas of various interested parties have yet to converge. In addition, RFID technology is changing at a rapid pace. Therefore, potential users of RFID technology face a number of uncertainties. However, potential users and others curious about this emerging technology, including the following, will find many of their concerns addressed in this book:

- Corporate decision makers who have received an RFID mandate from a customer or who want to adopt RFID for their enterprise
- IT managers who want to initiate a first RFID program
- Architects or developers who want to get practical tips and guidelines on implementing an RFID system and avoid the potential pitfalls
- Teachers who want to teach a course about RFID
- Students who want to know more about the technology to prepare for the RFID job market
- Consumers who want to be aware of how the technology is being used and its capabilities and limitations
- Anyone who is interested to know about the technology and its aspects

As you can see from the preceding list, this book provides hype-free information that addresses the concerns and needs of a wide audience. A good understanding of the various aspects of RFID can help potential users understand how the technology can be used to serve one's interests. Such an understanding can, in turn, potentially accelerate the adoption of RFID in everyday scenarios. This book provides a solid foundation for understanding RFID technology, and serves as a catalyst for wider adoption of the technology in our lives.

Although every effort has been made to ensure that this book is error free, some errors might have crept into print. Readers can e-mail the author at ibmpress@pearsoned.com with comments regarding this book and suggestions to make it better. Businesses can contact the author at this e-mail to describe RFID deployments. Vendors that are interested in listing products and services in this book can contact the author at this same e-mail address.

Sandip Lahiri
2005

Acknowledgments

I could not have written this book without the support, cooperation, and encouragement from several people. First and foremost, I am deeply indebted to Doug Hunt, Vice President, Global Application Innovation Services, IBM Business Consulting Services (BCS). I also want to thank John Baker, Service Area Leader—Wireless Applications, for his active support during the writing of this book. John was also instrumental in providing several critical insights that were invaluable to the development of the content. Special thanks go to my colleague and team leader Naresh Malik, who acted as a guide and mentor.

I am also grateful to Wainwright Ballard for giving me the opportunity to craft the RFID Bootcamp training course, which was one of the motivating factors behind writing this book. My colleagues Christopher Madison and Brian Eccles were a constant source of encouragement, and I want to thank them for their support. Thanks are also due to John Dorn, my first manager at IBM, for introducing me to the world of RFID. I also want to say, "Thanks a lot" to the following managers and colleagues who have directly or indirectly contributed to this effort: Ranjit Balaram, Sandy L. Anderson, Rob Wiegmans, Hjalmar Halvorsen, Sriram Ramanathan, David Ritcey, Deepak Mahbubani, Matt Davis, Frederick Rowe, Subu Musti, Nicky Sandhu, Gerry Gudgel, Linda Rodgers, Gloria Adams, and Robyn Schwartz; I am also grateful to Soumya Basu, Prithwis Mukerjee, and Gopal Bhageria.

I am extremely thankful to the vendors who have graciously supported me in this endeavor. I am truly indebted to all the reviewers of this book, with Wayne Steeves deserving a special mention for his detailed criticism of the content. Huge thanks go to the absolutely greatest editorial, marketing, and production teams at Prentice Hall PTR, who provided unflinching support for this book.

Last but certainly not least, I want to thank my family and friends for their wholehearted support during the writing of this book. My wife, Anasua, and daughters, Sohini and Aishani, cheerfully put up with my long working hours in the evenings and on weekends; my in-laws, Jiban and Abha Chatterjee, provided wonderful hospitality; and my pal, Prasenjit Pal (pun unintended), eagerly helped me out on several book-related issues when I was working on this book. Finally, I sincerely apologize if I have inadvertently neglected to mention someone who rightly deserves a thank you here.

Technology Overview

Radio frequency identification (RFID) technology uses radio waves to automatically identify physical objects (either living beings or inanimate items). Therefore, the range of objects identifiable using RFID includes virtually everything on this planet (and beyond). Thus, RFID is an example of *automatic identification* (Auto-ID) technology by which a physical object can be identified automatically. Other examples of Auto-ID include bar code, biometric (for example, using fingerprint and retina scan), voice identification, and *optical character recognition* (OCR) systems.

Consider the word *identify* more closely. Although two cans, A and B, of a particular brand of motor oil in a store might look identical, substantial differences between the two might in fact exist. For example,

- The retailer might have used two different order numbers to obtain cans A and B from the distributor.
- Can A might have been produced in North America, whereas can B might have been manufactured in Asia.
- A person named Bob might have loaded A onto the delivery truck, whereas a person named Chi might have loaded B onto a similar truck.
- Can A might have arrived in the store on a different date than when can B arrived.

Generally, although none of the preceding information appears on cans A or B for a person to view in a store, this information is nonetheless associated with these cans. You can, by using a set of such information, uniquely identify can A from can B. Also, even assuming that no such information exists, the very fact that that two distinct physical objects exist suggests the possibility to distinguish them (for example, by assigning a number that is unique to can A and one that is unique to can B). In summary, although cans A and B might look identical in appearance, composition, expiration date, recycling information, and so on, they can actually be differentiated in some way so that cans A and B, and any other can of motor oil produced by this particular manufacturer (or any other manufacturer), are *unique in some way*. When used in the context of RFID, the word *identify* refers to this uniqueness of an object.

The implications regarding object identity are tremendous. For example, consider how the preceding example of motor oil can be extended to other objects, irrespective of whether RFID technology can be used with:

- Every grain of rice consumed annually worldwide
- Every grain of sand on every beach worldwide
- Every leaf on every tree worldwide
- Every drop of rain that falls worldwide in a given year

The objects in this preceding list represent *potential* identification scenarios. Current RFID technology cannot be used to identify these objects. Even with technological advances (over the next 10 years, for example), some (or all) of these identification scenarios are unlikely. After all, how can you tag a raindrop, which has an extremely short life and dynamic behavior (such as dividing into smaller raindrops when it grows beyond 5 mm in size)?

Before delving into a detailed discussion of RFID technology, you need to understand the fundamental terms and concepts associated with RFID. The following section serves as an RFID technology primer.

1.1   Fundamental Concepts

A *wave* is a disturbance that transports energy from one point to another.

Electromagnetic waves are created by electrons in motion and consist of oscillating electric and magnetic fields. These waves can pass through a number of different material types.

The highest point of a wave is called a *crest*, and the lowest point is called a *trough*.

The distance between two consecutive crests or two consecutive troughs is called the *wavelength*.

One complete wavelength of oscillation of a wave is called a *cycle*.

The time taken by a wave to complete one cycle is called its *period of oscillation*.

The number of cycles in a second is called the *frequency* of the wave. The frequency of a wave is measured in *hertz* (abbreviated as Hz) and named in honor of the German physicist Heinrich Rudolf Hertz. If the frequency of a wave is 1 Hz, it means that the wave is oscillating at the rate of one cycle per second. It is common to express frequency in KHz (or kilohertz = 1,000 Hz), MHz (or megahertz = 1,000,000 Hz), or GHz (or gigahertz = 1,000,000,000 Hz).

Amplitude is the height of a crest or the depth of a trough from the undisturbed position. The former is also called the *positive amplitude*, and the latter the *negative amplitude*. In general, the *amplitude at a certain point* of a wave is its height or depth from the undisturbed position, and is called positive or negative accordingly.

Figure 1-1 shows several parts of a wave.

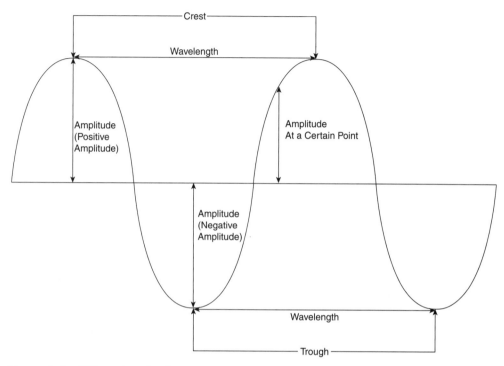

Figure 1-1   Different parts of a wave.

Radio or *radio frequency* (RF) waves are electromagnetic waves with wavelengths between 0.1 cm and 1,000 km. Another equivalent definition in terms of frequency is radio waves are electromagnetic waves whose frequencies lie between 30 Hz and 300 GHz. Other electromagnetic wave types are infrared, visible light wave, ultraviolet, gamma-ray, x-ray, and cosmic-ray.

RFID uses radio waves that are generally between the frequencies of 30 KHz and 5.8 GHz.

A *continuous wave* (CW) is a radio wave with constant frequency and amplitude. From a communications vantage, a CW does not have any embedded information in it but can be modulated to transmit a signal.

Modulation refers to the process of changing the characteristics of a radio wave to encode some information-bearing signal. Modulation can also refer to the result of applying the modulation process to a radio wave.

Radio waves can be affected by the material through which they propagate. A material is called *RF-lucent* or *RF-friendly* for a certain frequency if it lets radio waves at this frequency pass through it without any substantial loss of energy. A material is called *RF-opaque* if it blocks, reflects, and scatters RF waves. A material can allow the radio waves to propagate through it but with substantial loss of energy. These types of materials are referred to as *RF-absorbent*. The RF-absorbent or RF-opaque property of a material is *relative*, because it depends on the frequency. That is, a material that is RF-opaque at a certain frequency could be RF-lucent at a different frequency. The RF properties of some example materials are provided in Table 1-2, following a discussion of RFID frequency types.

Classes of RFID frequency types include the following:

- Low frequency (LF)
- High frequency (HF)
- Ultra high frequency (UHF)
- Microwave frequency

The following subsections discuss these frequency types.

1.1.1 Low Frequency (LF)

Frequencies between 30 KHz and 300 KHz are considered low, and RFID systems commonly use the 125 KHz to 134 KHz frequency range. A typical LF RFID system operates at 125 KHz or 134.2 KHz. RFID systems operating at LF generally use passive tags (discussed in Section 1.2.1), have low data-transfer rates from the tag to the reader, and are especially good if the operating environment contains metals, liquids, dirt, snow, or mud (a very important characteristic of LF systems). Active LF tags (discussed in Section 1.2.1) are also available from vendors. Because of the maturity of this type of tag, LF tag systems probably have the largest installed base. The LF range is accepted worldwide.

1.1.2 High Frequency (HF)

HF ranges from 3 MHz to 30 MHz, with 13.56 MHz being the typical frequency used for HF RFID systems. A typical HF RFID system uses passive tags, has a slow data-transfer rate from the tag to the reader, and offers fair performance in the presence of metals and liquids. HF systems are also widely used, especially in hospitals (where it does not interfere with the existing equipment). The HF frequency range is accepted worldwide.

The next frequency range is called *very high frequency* (VHF) and lies between 30 and 300 MHz. Unfortunately, none of the current RFID systems operate in this range. Therefore, this frequency type is not discussed any further.

1.1.3 Ultra High Frequency (UHF)

UHF ranges from 300 MHz to 1 GHz. A typical passive UHF RFID system operates at 915 MHz in the United States and at 868 MHz in Europe. A typical active UHF RFID system operates at 315 MHz and 433 MHz. A UHF system can therefore use both active and passive tags and has a fast data-transfer rate between the tag and the reader, but performs poorly in the presence of metals and liquids (*not* true, however, in the cases of low UHF frequencies such as 315 MHz and 433 MHz). UHF RFID systems have started being deployed widely because of the recent RFID mandates of several large private and public enterprises, such as several international and national retailers, the U.S. Department of Defense, and so on (see Chapter 10, "Standards"). The UHF range is *not* accepted worldwide.

1.1.4 Microwave Frequency

Microwave frequency ranges upward from 1 GHz. A typical microwave RFID system operates either at 2.45 GHz or 5.8 GHz, although the former is more common, can use both semi-active and passive tags, has the fastest data-transfer rate between the tag and the reader, and performs very poorly in the presence of metals and liquids. Because antenna length is inversely proportional to the frequency (see Section 1.2.1.1.2), the antenna of a passive tag operating in the microwave range has the smallest length (which results in a small tag size because the tag microchip can also be made very small). The 2.4 GHz frequency range is called *Industry, Scientific, and Medical* (ISM) band and is accepted worldwide.

International restrictions apply to the frequencies that RFID can use. Therefore, some of the previously discussed frequencies might not be valid worldwide. Table 1-1 lists some example frequency-use restrictions for RFID together with the maximum allowable *power* and *duty cycle* (explained later in this chapter).

Table 1-1 International RFID Frequency Regulations

Country/ Region	LF	HF	UHF	Microwave
United States	125–134 KHz	13.56 MHz 10 watts effective radiated power (ERP)	902-928 MHz, 1 watt ERP or 4 watts ERP with a directional antenna with at least 50-channel hopping.	2400–2483.5 MHz, 4 watts, ERP 5725–5850 MHz, 4 watts ERP
Europe	125–134 KHz	13.56 MHz	865–865.5 MHz, 0.1 watts ERP, Listen Before Talk (LBT). 865.6–867.6 MHz, 2 watts ERP, LBT. 867.6–868 MHz, 0.5 watts ERP, LBT.	2.45 GHz
Japan	125–134 KHz	13.56 MHz	Not allowed. MPHPT (Ministry of Public Management, Home Affairs, Posts and Telecommunications) has opened up 950–956 MHz band for experimentation.	2.45 GHz
Singapore	125–134 KHz	13.56 MHz	923–925 MHz. 2 watts ERP.	2.45 GHz
China	125–134 KHz	13.56 MHz	Not allowed. Future possibility: 840–843 MHz and/or 917-925 MHz. SAC (Standardization Administration of China) is entrusted to formulate the RFID regulations.	2446–2454 MHz, 0.5 watts ERP

Table 1-2 lists RF properties of some example materials.

Table 1-2   RF Properties of Example Material Types

Material	LF	HF	UHF	Microwave
Clothing	RF-lucent	RF-lucent	RF-lucent	RF-lucent
Dry wood	RF-lucent	RF-lucent	RF-lucent	RF-absorbent
Graphite	RF-lucent	RF-lucent	RF-opaque	RF-opaque
Liquids (some types)	RF-lucent	RF-lucent	RF-absorbent	RF-absorbent
Metals	RF-lucent	RF-lucent	RF-opaque	RF-opaque
Motor oil	RF-lucent	RF-lucent	RF-lucent	RF-lucent
Paper products	RF-lucent	RF-lucent	RF-lucent	RF-lucent
Plastics (some types)	RF-lucent	RF-lucent	RF-lucent	RF-lucent
Shampoo	RF-lucent	RF-lucent	RF-absorbent	RF-absorbent
Water	RF-lucent	RF-lucent	RF-absorbent	RF-absorbent
Wet wood	RF-lucent	RF-lucent	RF-absorbent	RF-absorbent

Radio waves are susceptible to interference from various sources, such as the following:

- Weather conditions such as rain, snow, and other types of precipitation. However, as mentioned before, these are not an issue at LF and HF.
- The presence of other radio sources such as cell phones, mobile radios, and so on.
- Electrostatic discharge (ESD). ESD is a sudden flow of electrical current through a material that is an insulator under normal circumstances. If a large potential difference exists between the two points on the material, the atoms between these two points can become charged and conduct electric current.

The discussion now turns to how RFID technology works.

A radio device called a *tag* is attached to the object that needs to be identified. Unique identification data about this tagged object is stored on this tag. When such a tagged object is presented in front of a suitable RFID reader, the tag transmits this data to the reader (via the reader antenna). The reader then reads the data and has the capability to forward it over suitable communication channels, such as a network or a serial connection, to a software application running on a computer. This application can then use this unique data to identify the object presented to the reader. It can then perform a variety of actions such as updating the location information of this object in the database, sending an alert to the floor personnel, or completely ignoring it (if a duplicate read, for example).

As you can understand from this description, RFID is also a data-collection technology. However, this technology has some unique characteristics that enable users to apply it in areas beyond the reach of traditional data-collection technologies, such as bar codes.

An RFID application is implemented by an RFID system, which constitutes the entire technology end-to-end.

1.2 RFID System

An *RFID system* is an integrated collection of components that implement an RFID solution.

An RFID system consists of the following components (in singular form) from an *end-to-end* perspective:

- **Tag.** This is a mandatory component of any RFID system.
- **Reader.** This is a mandatory component, too.
- **Reader antenna.** This is another mandatory component. Some current readers available today have built-in antennas.
- **Controller.** This is a mandatory component. However, most of the new-generation readers have this component built in to them.
- **Sensor, actuator, and annunciator.** These optional components are needed for external input and output of the system.
- **Host and software system.** Theoretically, an RFID system can function independently without this component. Practically, an RFID system is close to worthless without this component.
- **Communication infrastructure.** This mandatory component is a collection of both wired and wireless network and serial connection infrastructure needed to connect the previously listed components together to effectively communicate with each other.

Figure 1-2 is a schematic diagram of an RFID system. Figure 1-3 shows an instantiation of this schematic with example components.

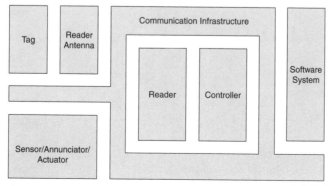

Figure 1-2   A schematic diagram of an RFID system.

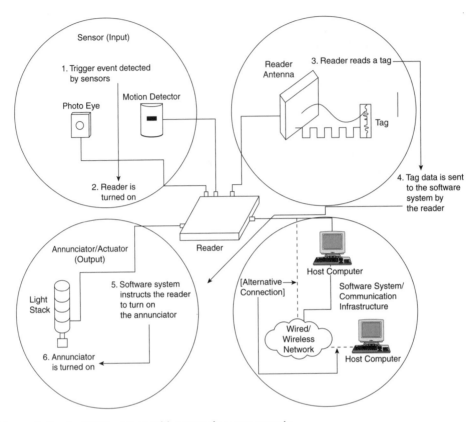

Figure 1-3   An RFID system with example components.

These figures may seem "reader-centric" because the RFID reader seems to be at the center of the entire system. Therefore, this figure might seem to be slanted, for example, toward the RFID vendor viewpoint. Figure 1-4 shows another perspective of the same system.

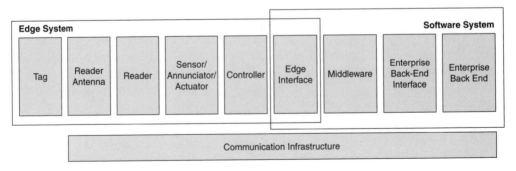

Figure 1-4   An RFID system from an IT perspective.

Note that in this scheme, the reader (together with the tag and antenna) is located at the *edge* of the system. This figure might be interpreted as how an RFID system looks from an IT or system-integrator perspective.

An RFID system thus has two parts—the first part (edge) governed by laws of physics and the second part involving information technology (IT). Which one is more important? The correct answer is "both." A state-of-the-art IT system is worthless if the data from its physical counterpart is unreliable and patchy. Similarly, a finely tuned RFID hardware setup is useless if the associated IT system cannot intelligently manage and process the data generated by this system.

An RFID system supports bidirectional communication flows, from the readers to the back end and from the back end to the readers (as also shown in Figure 1-3).

The following subsections discuss these previously identified RFID system components in detail.

1.2.1   Tag

An RFID *tag* is a device that can store and transmit data to a reader in a contactless manner using radio waves.

RFID tags can be classified in two different ways. The following list shows the first classification, which is based on whether the tag contains an on-board power supply and/or provides support for specialized tasks:

- Passive
- Active
- Semi-active (also known as semi-passive)

The following subsections discuss these in detail. (The other classification is discussed after this.)

1.2.1.1   Passive Tags

This type of RFID tag does not have an on-board power source (for example, a battery), and instead uses the power emitted from the reader to energize itself and transmit its stored data to the reader. A passive tag is simple in its construction and has no moving parts. As a result, such a tag has a long life and is generally resistant to harsh environmental conditions. For example, some passive tags can withstand corrosive chemicals such as acid, temperatures of 400°F (204°C approximately), and more.

In tag-to-reader communication for this type of tag, a reader always communicates first, followed by the tag. The presence of a reader is mandatory for such a tag to transmit its data.

A passive tag is typically smaller than an active or semi-active tag. It has a variety of read ranges starting with less than 1 inch to about 30 feet (9 meters approximately).

A passive tag is also generally cheaper compared to an active or semi-active tag.

A *contactless smart card* is a special type of passive RFID tag that is widely used today in various areas (for example, as ID badges in security and loyalty cards in retail). The data on this card is read when it is in close proximity to a reader. The card does not need to be physically in contact with the reader for reading.

A passive tag consists of the following main components:

- Microchip
- Antenna

Figure 1-5 shows the components of a passive tag.

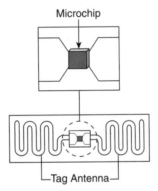

Figure 1-5   Components of a passive tag.

The following subsections discuss these components in detail.

1.2.1.1.1   Microchip

Figure 1-6 shows the basic components of a microchip.

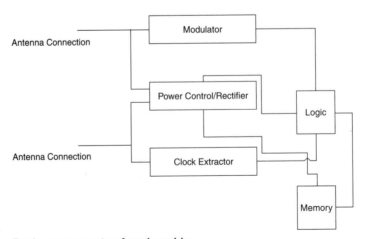

Figure 1-6   Basic components of a microchip.

The *power control/rectifier* converts AC power from the reader antenna signal to DC power. It supplies power to the other components of the microchip. The *clock extractor* extracts the clock signal from reader antenna signal. The *modulator* modulates the received reader signal. The tag's response is embedded in the modulated signal, which is then transmitted back to the reader. The *logic* unit is responsible for implementing the communication protocol between the tag and the reader. The microchip *memory* is used for storing data. This memory is generally segmented (that is, consists of several blocks or fields). *Addressability* means the ability to address (that is, read or write) the individual memory of a tag's microchip. A tag memory block can hold different data types, such as a portion of the tagged object identifier data, checksum (for example, cyclic redundancy check [CRC]) bits for checking the accuracy of the transmitted data, and so on. Recent advances in technology have shrunk the size of the microchip to less than the size of a grain of sand. However, a tag's physical dimensions are not determined by the size of its microchip but by the length of its antenna.

1.2.1.1.2   Antennas

A tag's antenna is used for drawing energy from the reader's signal to energize the tag and for sending and receiving data from the reader. This antenna is physically attached to the microchip. The antenna geometry is central to the tag's operations. Infinite variations of antenna designs are possible, especially for UHF, and designing an effective antenna for a tag is as much as an art as a science. The antenna length is directly proportional to the tag's operating wavelength. A *dipole* antenna consists of a straight electric conductor (for example, copper) that is interrupted at the center. The total length of a dipole antenna is half the wavelength of the used frequency to optimize the energy transfer from the reader antenna signal to the tag. A *dual dipole* antenna consists of two dipoles, which can greatly reduce the tag's alignment sensitivity. As a result, a reader can read this tag at different tag orientations. A *folded dipole* consists of two or more straight electric conductors connected in parallel and each half the wavelength (of the used frequency) long. When two conductors are involved, the resulting folded dipole is called *2-wire folded dipole*. A *3-wire folded dipole* consists of three conductors connected in parallel. Figure 1-7 shows these antenna types.

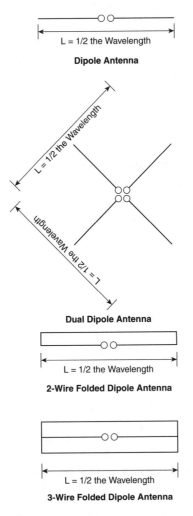

Figure 1-7 Dipole antenna types.

A tag's antenna length is generally much larger than the tag's microchip, and therefore ultimately determines a tag's physical dimensions. An antenna can be designed based on several factors, such as the following:

- Reading distance of the tag from the reader
- Known orientation of the tag to the reader
- Arbitrary orientation of the tag to the reader
- Particular product type(s)
- Speed of the tagged object
- Specific operating condition(s)
- Reader antenna polarization

The connection points between a tag's microchip and the antenna are the weakest links of the tag. If any of these connection points are damaged, the tag might become nonfunctional or might have its performance significantly degraded. An antenna designed for a specific task (such as tagging a case) might perform poorly for a different task (such as tagging an individual item in the case). Changing antenna geometry randomly (just "hacking around;" for example, cutting or folding it) is not a good idea because this can detune the tag, resulting in suboptimal performance. However, someone who knows what he is doing can deliberately modify a tag's antenna to detune it (drilling a hole into it, for example) and actually increase the readability of the tag!

Currently, a tag antenna is constructed with a thin strip of a metal (for example, copper, silver, or aluminum). In the future, however, it will be possible to print antennas directly on the tag label, case, and product packaging using a conductive ink that contains copper, carbon, or nickel. Effort is also currently underway to determine whether the microchip might be printed with such an ink, too. These future enhancements may enable you to print an RFID tag just as you do a bar code on the case and item packaging. As a result, the cost of an RFID tag might drop substantially below the anticipated $.05 per tag. Even without the ability to print a microchip, a printed antenna can be attached to a microchip to create a complete RFID tag much faster than attaching a metal antenna.

Figures 1-8 through 1-10 show passive tags from various vendors.

Figure 1-8   Family of LF tags from Texas Instruments.

Reprinted with permission from Texas Instruments

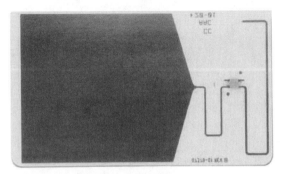

Figure 1-9   915 MHz tag from Intermec Corporation.

Reprinted with permission from Intermec Technologies Corporation

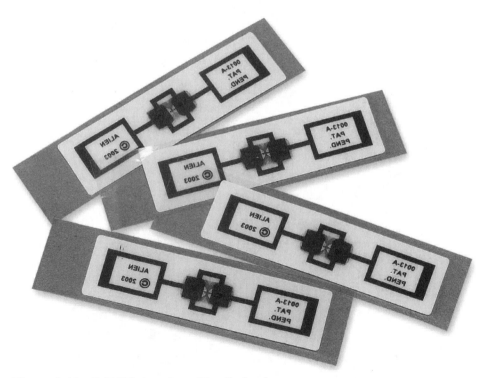

Figure 1-10   2.45 GHz tags from Alien Technology.

Reprinted with permission from Alien Technology

1.2.1.2   Active Tags

Active RFID tags have an on-board power source (for example, a battery; other sources of power, such as solar, are also possible) and electronics for performing specialized tasks. An active tag uses its on-board power supply to transmit its data to a reader. It does not need the reader's emitted power for data transmission. The on-board electronics can contain microprocessors, sensors, and input/output ports powered by the on-board power source. Therefore, for example, these components can measure the surrounding temperature and generate the average temperature data. The components can then use this data to determine other parameters such as the expiry date of the attached item. The tag can then transmit this information to a reader (along with its unique identifier). You can think of an active tag as a wireless computer with additional properties (for example, like that of a sensor or a set of sensors).

In tag-to-reader communication for this type of tag, a tag always communicates first, followed by the reader. Because the presence of a reader is not necessary for data transmission, an active tag can broadcast its data to its surroundings even in the absence of a reader. This type of active tag, which continuously transmits data with or without the presence of a reader, is also called a *transmitter*. Another type of active tag enters a sleep or a low-power state in the absence of interrogation by a reader. A reader wakes up such a tag from its sleep state by issuing an appropriate command. This state saves the battery power, and therefore, a tag of this type generally has a longer life compared to an active transmitter tag. In addition, because the tag transmits only when interrogated, the amount of induced RF noise in its environment is reduced. This type of active tag is called a *transmitter/receiver* (or a *transponder*). As you can understand from this discussion, you cannot accurately call all tags transponders.

The reading distance of an active tag can be 100 feet (30.5 meters approximately) or more when the active transmitter of such a tag is used.

An active tag consists of the following main components:

- **Microchip.** The microprocessor size and capabilities are generally greater than the microchips found in passive tags.
- **Antenna.** This can be in the form of an RF module that can transmit the tag's signals and receive reader's signals in response. For a semi-active tag, this is composed of thin strip(s) of metal such as copper, similar to that of a passive tag.
- **On-board power supply.**
- **On-board electronics.**

Figure 1-11 shows examples of active and semi-active tags.

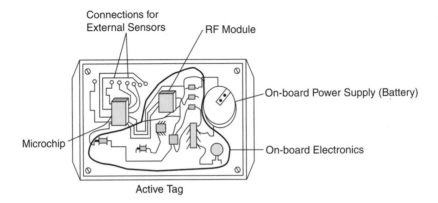

Active Tag

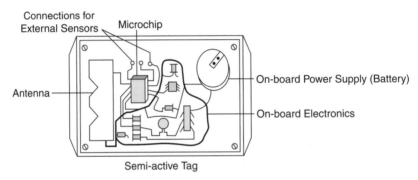

Semi-active Tag

Figure 1-11   Example active and semi-active tags.

The first two components have already been described in the previous section. The last two components are discussed now.

1.2.1.2.1   On-Board Power Supply

All active tags carry an on-board power supply (for example, a battery) to provide power to its on-board electronics and to transmit data. If a battery is used, an active tag generally lasts for about 2 to 7 years depending on the battery life. One of the determining factors of the battery life is the data-transmission rate interval of the tag—the larger the interval, the longer the battery and hence the tag life. For example, suppose that an active tag is made to transmit once every few seconds. If you increase this so that the tag transmits once every few minutes or even once every few hours, you extend the battery life. The on-board sensors and processors consume power and can shorten the battery life, too.

When the battery of an active tag is completely discharged, the tag stops transmitting messages. A reader that was reading these messages does not know whether the tag's battery has died or whether the tagged product has disappeared from its read zone unless the tag transmits its battery status to this reader.

1.2.1.2.2   On-Board Electronics

The on-board electronics allow the tag to act as a transmitter, and optionally allow it to perform specialized tasks such as computing, displaying the values of certain dynamic parameters, acting as a sensor, and so on. This component can also provide an option for connecting external sensors. Therefore, depending on the sensor type attached, such a tag can perform a wide variety of sensing tasks. In other words, the range of functionality of this component is virtually limitless. Note that as the functionality and hence the physical size of this component grows, the tag might grow in size. This growth is acceptable because no hard limit applies to the size of an active tag as long as it can be deployed (that is, properly attached to the object that needs to be tagged). This means active tags can be applied to a wide range of applications, several of which might not even exist today.

1.2.1.3   Semi-Active (Semi-Passive) Tags

Semi-active tags have an on-board power source (for example, a battery) and electronics for performing specialized tasks. The on-board power supply provides energy to the tag for its operation. However, for transmitting its data, a semi-active tag uses the reader's emitted power. A semi-active tag is also called a *battery-assisted tag*. In tag-to-reader communication for this type of tag, a reader always communicates first, followed by the tag. Why use a semi-passive tag over a passive tag? Because a semi-active tag does not use the reader's signal, unlike a passive tag, to excite itself, it can be read from a longer distance as compared to a passive tag. Because no time is needed for energizing a semi-active tag, such a tag could be in the read zone of a reader for substantially less time for its proper reading (unlike a passive tag). Therefore, even if the tagged object is moving at a high speed, its tag data can still be read if a semi-active tag is used. Finally, a semi-active tag might offer better readability for tagging of RF-opaque and RF-absorbent materials. The presence of these materials might prevent a passive tag from being properly excited, resulting in failure to transmit its data. However, this is not an issue with a semi-active tag.

The reading distance of a semi-active tag can be 100 feet (30.5 meters approximately) under ideal conditions using a modulated backscatter scheme (in UHF and microwave).

Figures 1-12 through 1-14 show active and semi-active tags from various vendors.

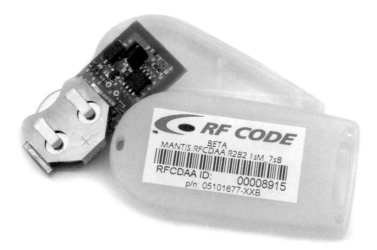

Figure 1-12   Mantis low UHF (303.8 MHz) active tag with built-in motion detector from RFCode, Inc.

Reprinted with permission from RFCode, Inc.

Figure 1-13   915 MHz/2.45 GHz semi-active tags from TransCore.

Reprinted with permission from TransCore

Figure 1-14   2.45 GHz semi-active tags from Alien Technology.

Reprinted with permission from Alien Technology

The next classification, as shown here, is based on the capability to support data rewrites:

- Read-only (RO)
- Write once, read many (WORM)
- Read-write (RW)

Both active and passive tags can be RO, WORM, and RW. The following sections discuss these classifications in detail.

1.2.1.4   Read Only (RO)

An RO tag can be programmed (that is, written) just once in its lifetime. The data can be burned into the tag at the factory during the manufacturing stage. To accomplish this, the individual fuses on the tag microchip are burned permanently using a fine-pointed laser beam. After this is done, the data cannot be rewritten for the entire lifetime of the tag. Such a tag is also called *factory programmed*. The tag manufacturer supplies the data on the tag, and the tag users typically do not have any control over it. This type of tag is good for small applications only, but is impractical for large manufacturing or when tag data needs to be customized based on the application. This tag type is used today in small pilots and business applications.

1.2.1.5   Write Once, Read Many (WORM)

A WORM tag can be programmed or written once, which is generally done not by the manufacturer but by the tag user right at the time when the tag needs to be created. In practice, however, because of buggy implementation, it is possible to overwrite particular types of WORM tag data several times (about 100 times is not uncommon)! If the data for such a tag is rewritten more than a certain number of times, the tag can be damaged permanently. A WORM tag is also called *field programmable*.

This type of tag offers a good price-to-performance ratio with reasonable data security, and is the most prevalent type of tag used in business today.

1.2.1.6   Read Write (RW)

An RW tag can be reprogrammed or rewritten a large number of times. Typically, this number varies between 10,000 and 100,000 times and above! This rewritability offers a tremendous advantage because the data can be written either by the readers or by the tag itself (in case of active tags). An RW tag typically contains a Flash or a FRAM memory device to store its data. An RW tag is also called *field programmable* or *reprogrammable*. Data security is a challenge for RW tags. In addition, this type of tag is most expensive to produce. RW tags are not widely used in today's applications, a fact that might change in the future as the tag technology and applicability increases with a decrease in tag cost.

It is important to briefly pause here and describe a type of RFID tag called *surface acoustic wave* (SAW) before moving on to the next topic.

1.2.1.7   SAW (Surface Acoustic Wave) Tags

A SAW tag differs fundamentally from microchip-based tags. SAW tags have started appearing on the market, and might be widely used in the future. Currently, SAW devices are widely used in cell phones, color televisions, and so on.

SAW tags use low-power RF waves in the ISM 2.45 GHz frequency range for their operation. Unlike a microchip-based tag, a SAW tag does not need DC power to energize itself for data transmission. Figure 1-15 shows how such a tag operates.

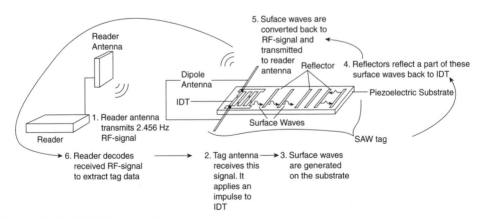

Figure 1-15   SAW tag operation.

A SAW tag consist of a dipole antenna attached to an *interdigital transducer* (IDT) placed on a piezoelectric substrate made of lithium niobate or lithium tantalate. A series of well-placed individual electrodes acting as reflectors (made of aluminum or etched on the substrate) are positioned on the substrate. The antenna applies an electrical impulse to the IDT when it receives the RF signal from a SAW reader. This impulse generates surface waves, also known as *Raleigh* waves, typically traveling at 3,000 to 4,000 meters per second on the substrate. Some of these waves are reflected back to the IDT by the reflectors; the rest are absorbed by the substrate. The reflected waves form a unique pattern, determined by the reflector positions, representing the tag data. These waves are converted back to the RF signal in the IDT and transmitted back to the RFID reader via the tag antenna. The reader then decodes the received signal to extract the tag data.

The advantages of a SAW tag include the following:

- Uses very low power because it does not need a DC source of power to energize itself.
- Can successfully tag RF-opaque and RF-absorbent materials, such as metal and water, respectively.
- Has a longer read range compared to a microchip tag operating in the same frequency range (that is, 2.45 GHz).
- Can operate with short bursts of RF-signal in contrast to microchip-based tags, which need much longer signal duration from reader to the tag.
- Has high read accuracy rates.
- Is hardy because of its simple design.
- Does not need anti-collision protocols. Anti-collision protocols need to be implemented at the reader level only instead of at both reader and tag level as for a microchip tag (thus reducing the cost of a SAW tag).

SAW readers are less prone to interference with other SAW readers. SAW tags might very well be the only choice in certain tagging situations and are likely to be widely used in the future.

Some tags can transmit data to a reader without using RF waves. A brief description of such tags follows.

1.2.1.8 Non-RFID Tags

The concept of attaching a tag and having it wirelessly transmit its unique ID to a reader is not the exclusive domain of RF waves. You can use other types of wireless communications for this purpose. For example, you can use ultrasonic and infrared waves for tag-to-reader communication.

Ultrasonic communication has the additional advantages that it does not cause interference with existing electrical equipment and cannot penetrate through walls. As a result, ultrasonic tagging systems can be deployed in hospitals, where such technology can coexist with the existing medical equipment. In addition, an ultrasonic reader and a tag must be within the same room for the tag to be read by the reader. This required proximity can prove helpful in asset monitoring and tracking.

An infrared tag uses light to transmit its data to a reader. Because light cannot penetrate through walls, an infrared tag and reader must both be in the same room for communication. If an obstacle covers the light source of a tag, the tag can no longer communicate with a reader (a serious disadvantage).

1.2.2   Readers

An RFID reader, also called an *interrogator*, is a device that can read from and write data to compatible RFID tags. Thus, a reader also doubles up as a writer. The act of writing the tag data by a reader is called *creating* a tag. The process of creating a tag and uniquely associating it with an object is called *commissioning the tag*. Similarly, *decommissioning a tag* means to disassociate the tag from a tagged object and optionally destroy it. The time during which a reader can emit RF energy to read tags is called the duty cycle of the reader. International legal limits apply to reader duty cycles.

The reader is the central nervous system of the entire RFID hardware system—establishing communication with and control of this component is the most important task of any entity which seeks integration with this hardware entity.

A reader has the following main components:

- Transmitter
- Receiver
- Microprocessor
- Memory
- Input/output channels for external sensors, actuators, and annunciators (Although, strictly speaking, these are optional components, they are almost always provided with a commercial reader.)
- Controller (which may reside as an external component)
- Communication interface
- Power

Figure 1-16 shows an example reader with these components.

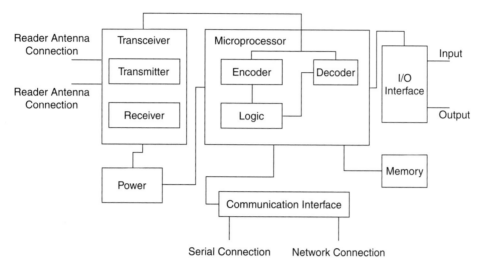

Figure 1-16   The components of an example reader.

The following subsections describe these components.

1.2.2.1   Transmitter

The reader's transmitter is used to transmit AC power and the clock cycle via its antennas to the tags in its read zone. This is a part of the *transceiver* unit, the component responsible for sending the reader's signal to the surrounding environment and receiving tag responses back via the reader antenna(s). The antenna ports of a reader are connected to its *transceiver* component. One reader antenna can be attached to each such antenna port. Currently, some readers can support up to four antenna ports.

1.2.2.2   Receiver

This component is also part of the *transceiver* module. It receives analog signals from the tag via the reader antenna. It then sends these signals to the reader microprocessor, where it is converted to its equivalent digital form (that is, the digital representation of the data that the tag has transmitted to the reader antenna).

1.2.2.3   Microprocessor

This component is responsible for implementing the reader protocol to communicate with compatible tags. It is performs decoding and error checking of the analog signal from the receiver. In addition, the microprocessor might contain custom logic for doing low-level filtering and processing of read tag data.

1.2.2.4   Memory

Memory is used for storing data such as the reader configuration parameters and a list of tag reads. Therefore, if the connection between the reader and the controller/software system goes down, not all read tag data will be lost. Depending on the memory size, however, a limit applies as to how many such tag reads can be stored at any one time. If the connection remains down for an extended period with the reader reading tags during this downtime, this limit might be exceeded and part of the stored data lost (that is, overwritten by the other tags that are read later).

1.2.2.5   Input/Output Channels for External Sensors, Actuators, and Annunciators

Readers do not have to be turned on for reading tags at all times. After all, the tags might appear only at certain times in the read zone, and leaving readers perpetually on would just waste the reader's energy. In addition, as mentioned previously, regulatory limits apply to the reader duty cycle, too. This component provides a mechanism for turning a reader on and off depending on external events. A sensor of some sort, such as a motion or light sensor, detects the presence of tagged objects in the reader's read zone. This sensor can then set the reader on to read this tag. Similarly, this component also allows the reader to provide local output depending on some condition via an annunciator (for example, sounding an audible alarm) or an actuator (for example, opening or closing a security gate, moving a robot arm, and so forth). Sensors, actuators, and annunciators are discussed later in this chapter.

1.2.2.6   Controller

A *controller* is an entity that allows an external entity, either a human or a computer program, to communicate with and control a reader's functions and to control annunciators and actuators associated with this reader. Often, manufacturers integrate this component into the reader itself (as firmware, for example). However, it is also possible to package this as a separate hardware/software component that must be bought together with the reader. Controllers are discussed in detail later in this chapter.

1.2.2.7   Communication Interface

The communication interface component provides the communication instructions to a reader that allow it to interact with external entities, via a controller, to transfer its stored data and to accept commands and send back the corresponding responses. You can assume that this interface component is either part of the controller or is the medium that lies between a controller and the external entities. This entity has important characteristics that make it necessary to treat this as an independent component. A reader could have a serial as well as a network interface for communication. A serial interface is probably the most widespread type of reader interface available, but next-generation readers are being developed with network interfaces as a standard feature. Sophisticated readers offer features such as automatic discovery by an application, embedded Web servers that allow the reader to accept commands and display the results using a standard Web browser, and so forth.

1.2.2.8 Power

This component supplies power to the reader components. The power source is generally provided to this component through a power cord connected to an appropriate external electrical outlet.

Like tags, readers can also be classified using two different criteria. The first criterion is the interface that a reader provides for communication. Based on this, readers can be classified as follows:

- Serial
- Network

The following subsections describe these reader types.

1.2.2.9 Serial Reader

Serial readers use a serial communication link to communicate with an application. The reader is physically connected to a computer's serial port using an RS-232 or RS-485 serial connection. Both of these connections have an upper limit on the cable length that can be used to connect a reader to a computer. RS-485 allows a longer cable length than RS-232 does.

The advantage of serial readers is that the communication link is reliable compared to network readers. Therefore, the use of these readers is recommended to minimize dependency on a communication channel.

The disadvantage of serial readers is the dependence on the maximum length of cable that can be used to connect a reader to a computer. In addition, because the number of serial ports is generally limited on a host, a larger number of hosts (as compared to the number of hosts needed for network readers) might be needed to connect to all the serial readers. Another problem is maintenance—if the firmware needs to be updated, for example, maintenance personnel might have to physically deal with each reader. Also, the serial data-transmission rate is generally lower than the network data-transmission rate. These factors might result in higher maintenance costs and significant operation downtime.

1.2.2.10 Network Reader

Network readers can be connected to a computer using both wired and wireless networks. In effect, the reader behaves like a network device installation that does not require any specialized knowledge of the hardware. Note, however, that SNMP-type monitoring features are currently available for just a few network reader types. Therefore, the majority of these readers cannot be monitored as standard network devices.

The advantage of network readers is that there is no dependence on the maximum length of cable that can be used to connect a reader to a computer. A smaller number of hosts are generally needed as compared to the serial readers. In addition, the reader firmware can be updated remotely over the network without any need to visit the reader physically. This can ease the maintenance effort and lower the cost of ownership of such an RFID system.

The disadvantage of network readers is that the communication link is not as reliable compared to serial readers. When the communication link goes down, the back end cannot be accessed. As a result, the RFID system might come to a complete standstill. The readers, in general, have internal memory to store tag reads. This feature might somewhat alleviate short network outages.

The next classification of reader type can be made based on its mobility, as follows:

- Stationary
- Handheld

The following subsections describe these reader types.

1.2.2.11   Stationary Reader

A stationary reader, also called a *fixed* reader, is what its name implies. These readers are mounted on a wall, portal, or some suitable structure in the read zone. The structure on which the reader is mounted may not be static! For example, some stationary readers are mounted on forklifts. Similarly, you can mount these readers inside delivery trucks. In contrast to tags, readers are not generally very tolerant of harsh environmental conditions. Therefore, if you install a reader outdoors or on moving objects, take care to ruggedize it properly. Stationary readers generally need external antennas for reading tags. A reader can provide up to four external antennas ports.

The cost of a stationary reader is generally less than the cost of handheld readers. Stationary readers are the most common type of reader used today.

Figures 1-17 and 1-18 show some fixed readers.

Figure 1-17   UHF fixed network reader from Alien Technology.

Reprinted with permission from Alien Tecnology

Figure 1-18   Low UHF (303.8 MHz) fixed wired/wireless (802.11b) network reader from RFCode, Inc.

Reprinted with permission from RFCode, Inc.

A type of reader called an *agile reader* can operate in different frequencies or can use different tag-to-reader communications protocols. Today's agile readers are generally stationary.

A type of stationary reader called an *RFID printer* can print a bar code and create (that is, write) an RFID tag on a *smart label* in an integrated operation. A smart label consists of a bar code label that has an embedded RFID tag in it. Various types of information, such as the sender and recipient addresses, product information, and free-form text, can be printed on the label, too. An RFID printer reads the smart label tag that it has just written to validate the write operation. If this validation fails, the printer rejects the smart label that it has just printed. This device obviates the necessity to separately create an RFID tag where bar codes are currently used (which might reduce additional logistics overhead). A business that is using bar codes today for its operations can use RFID printers as a first step in adopting the RFID technology. The bar code information provides a human-readable identification of the tagged object. Also, the existing systems and operations can keep using the same bar code data with some or no change. The notes area of the label can provide the embedded tag ID in human-readable form. The RFID tag can provide object-level Auto-ID capabilities and other associated benefits. Figure 1-19 shows an example smart label. Figure 1-20 shows an example RFID printer.

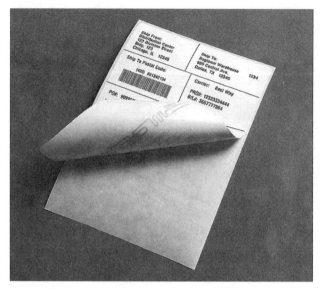

Figure 1-19   RFID smart label from Zebra Technologies.

Reprinted with permission from Zebra Technologies

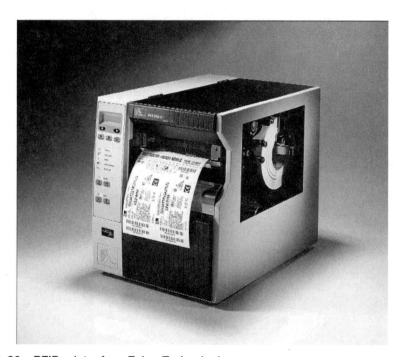

Figure 1-20   RFID printer from Zebra Technologies.

Reprinted with permission from Zebra Technologies

A stationary reader can generally operate in the following modes:

- Autonomous
- Interactive

The following subsections describe these modes.

1.2.2.11.1   Autonomous Mode

In autonomous mode, a reader continuously read tags in its read zone. Every time a tag is read, it is saved to a list, usually called a *tag list*. An item on the tag list is associated with what is generally called a *persist time*. If the associated tag cannot be read for a period of time exceeding its persist time, it is dropped from the tag list. An application running on a host machine can register itself to receive the tag list periodically. A tag list includes information such as the following:

- Unique tag identifiers
- Reading time
- How many times a particular tag has been read since it has been discovered (that is, first read by the reader)
- The antenna ID that read a particular tag
- Reader name

1.2.2.11.2   Interactive Mode

In interactive mode, a read receives and executes commands from an application running on a host machine or from a user using a vendor-supplied client to communicate with the reader. After the reader fully executes the current command, it waits for the next. A reader can execute a range of commands, from sending the current tag list to the command invoker to changing the reader's configuration parameters.

1.2.2.12   Handheld Reader

A handheld reader is a mobile reader that a user can operate as a handheld unit. A handheld reader generally has built-in antenna(s). Although these readers are typically the most expensive (and few are commercially available), recent advances in reader technology are resulting in sophisticated handheld readers at lower prices. Figures 1-21 shows a handheld reader.

Figure 1-21   UHF handheld reader from Intermec Corporation.

Reprinted with permission from Intermec Technologies Corporation

The following section introduces the underlying communication mechanisms between a tag and a reader.

1.2.2.13   Communication Between a Reader and a Tag

Depending on the tag type, the communication between a reader and a tag can be one of the following:

- Modulated backscatter

- Transmitter type

- Transponder type

Before delving into the details of these communication types, it is important for you to understand the concepts of near field and far field.

The area between a reader antenna and one full wavelength of the RF wave emitted by the antenna is called *near field*. The area beyond one full wavelength of the RF wave emitted from a reader antenna is called *far field*. Passive RFID systems operating in LF and HF use near field communication, whereas those in UHF and microwave frequencies use far field communication. The signal strength in near field communication attenuates as the cube of the distance from the reader antenna. In far field, it attenuates as square of the distance from the reader antenna. As a result, far field communication is associated with a longer read range compared with near field communication.

Next, a comparison between tag read and tag write is in order.

Tag write takes a longer time than tag read under the same conditions because a write operation consists of multiple additional steps, including an initial verification, erasing any existing tag data, writing the new tag data, and a final verification phase. In addition, the data is written on the tag in blocks in multiple steps. As a result, a single tag write can take hundreds of milliseconds to complete and increases with the increase in data size. In contrast, several tags can be read in this time interval by the same reader. Also, tag write is a sensitive process that needs the target tag to be closer (compared to its corresponding read distance) to the reader antenna for the entire write operation. This closer proximity ensures the tag antenna can derive sufficient energy from the reader antenna signal to power its microchip so that it can execute the write instructions. The power requirement for write operation is generally significantly higher than that required for reading. The write operation might fail otherwise. However, a tag does not have to stay close to the reader during a read operation. Also, during tag write operation, any tag other than the target should not be in write range of the reader. Otherwise, in some cases, this other tag might accidentally get written rather than the target tag. This write range issue is clearly not relevant during a read operation, when multiple tags can exist in the read range of the reader at the same time.

1.2.2.13.1 Modulated Backscatter

Modulated backscatter communication applies to passive as well as to semi-active tags. In this type of communication, the reader sends out a *continuous wave* (CW) RF signal containing AC power and clock signal to the tag at the *carrier frequency* (the frequency at which the reader operates). Through physical coupling (that is, a mechanism by which the transfer of energy takes place from the reader to the tag), the tag antenna supplies power to the microchip. The word excite is frequently used to indicate a passive tag microchip drawing power from a reader's signal to properly energize itself. About 1.2 volts are generally necessary to energize the tag microchip for reading purposes. For writing, the microchip usually needs to draw about 2.2 volts from the reader signal. The microchip now modulates or breaks up the input signal into a sequence of on and off patterns that represents its data and transmits it back. When the reader receives this modulated signal, it decodes the pattern and obtains the tag data.

Thus, in modulated backscatter communication, the reader always "talks" first, followed by the tag. A tag using this scheme cannot communicate at all in the absence of a reader because it depends totally on the reader's power to transmit its data. Figure 1-22 shows backscatter communication.

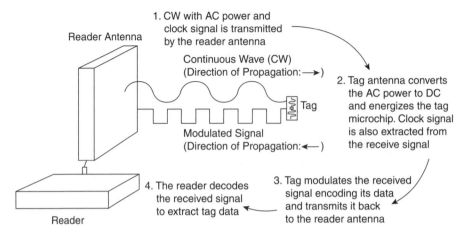

Figure 1-22   Backscatter communication.

A related term, *beam power*, is also used in this context, and means that a tag is using the reader's power to modulate the reader signal back. Note that a passive tag exclusively uses beam power to transmit its data. A semi-active tag uses beam power to clock its oscillator and generate the transmit signal back. Thus, in essence, a semi-active tag also uses beam power to transmit its data.

1.2.2.13.2   Transmitter Type

This type of communication applies to active tags only. In this type of communication, the tag broadcasts its message to the environment in regular intervals, irrespective of the presence or absence of a reader. Therefore, in this type of communication, the tag always "talks" first rather than the reader. Figure 1-23 shows transmitter communication.

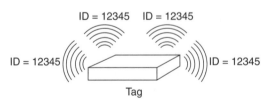

Figure 1-23   Transmitter communication.

1.2.2.13.3   Transponder Type

This type of communication applies to a special type of active tags called transponders (as discussed previously). In this type of communication, the tag goes to a "sleep" or into a dormant stage in the absence of interrogation from a reader. In this state, the tag might periodically send a message to check whether any reader is listening to it. When a reader receives such a query message, it can instruct the tag to "wake up" or end the dormant state. When the tag receives this command from the reader, it exits its current state and starts to act as a transmitter tag again. (That is, it starts broadcasting its message periodically to its surroundings.) In this type of communication, the tag data is sent only when the reader specifically asks for it. Figure 1-24 shows transponder communication.

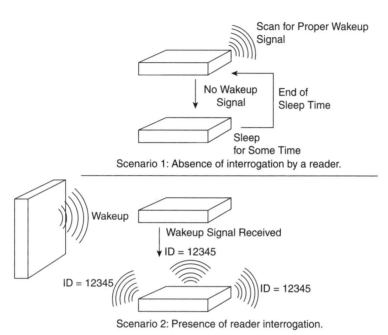

Scenario 1: Absence of interrogation by a reader.

Scenario 2: Presence of reader interrogation.

Figure 1-24 Transponder communication.

1.2.3 Reader Antenna

A reader communicates to a tag through the reader's antennas, a separate device that is physically attached to a reader, at one of its antenna ports, by means of a cable. This cable length is generally limited to between 6 and 25 feet. (However, this length limit may vary.) As mentioned previously, a single reader can support up to four antennas (that is, have four physical antenna ports). A reader antenna is also called the reader's *coupling element* because it creates an electromagnetic field to *couple* with the tag. An antenna broadcasts the reader transmitter's RF signal into its surroundings and receives tag responses on the reader's behalf. Therefore, proper positioning of the antennas, *not the readers*, is essential for good read accuracy (although a reader has to be located somewhat close to an antenna because of the limitation of the antenna cable length). In addition, some stationary readers might have in-built antennas. As a result, in this case, positioning the antennas for a reader is equivalent to positioning the reader itself. In general, RFID reader antennas are shaped like rectangular or square boxes. Figures 1-25 and 1-26 show some reader antennas.

Figure 1-25   UHF Circular polarized reader antenna from Alien Technology.

Reprinted with permission from Alien Technology

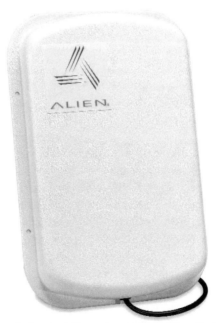

Figure 1-26   UHF Linear polarized reader antenna from Alien Technology.

Reprinted with permission from Alien Technology

It is now time to discuss a very important concept of an antenna: the antenna footprint.

1.2.3.1   Antenna Footprint

The footprints of the reader's antennas determine the read zone (also called the *read window*) of a reader. In general, an antenna footprint, also called an *antenna pattern*, is a three-dimensional region shaped somewhat like an ellipsoid or a balloon projecting out of the front of the antenna. In this region, the antenna's energy is most effective; therefore, a reader can read a tag placed inside this region with the least difficulty. Figure 1-27 shows such a simple antenna pattern.

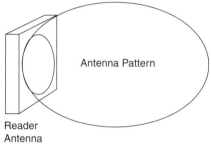

Figure 1-27   Simple antenna pattern.

In reality, because of antenna characteristics, the footprint of an antenna is never uniformly shaped like an ellipsoid but almost always contains deformities or protrusions. Each protrusion is surrounded by dead zones. Such dead zones are also called *nulls*. Figure 1-28 shows an example of such an antenna pattern.

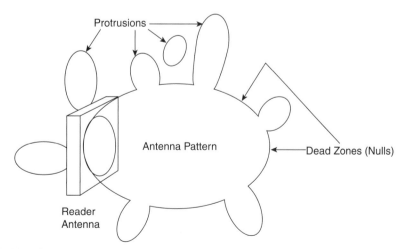

Figure 1-28   An example antenna pattern containing protrusions.

The reflection of reader antenna signals on RF-opaque objects causes what is known as *multipath*. In this case, the reflected RF waves are scattered and can arrive at the reader antenna at different times using different paths. Some of the arriving waves could be *in phase* (that is, exactly match with the original antenna signal's wave pattern). In this case, the original antenna signal is enhanced when these waves impose with the original waves giving rise to protrusions. This phenomenon is also known as *constructive interference*. Some of the waves could also arrive out of phase (that is, the exact opposite of the original antenna wave pattern). In this case, the original antenna signal is cancelled when these two wave types impose on each other. This is also called *destructive interference*. Nulls are created as a result. Figure 1-29 shows an example of multipath.

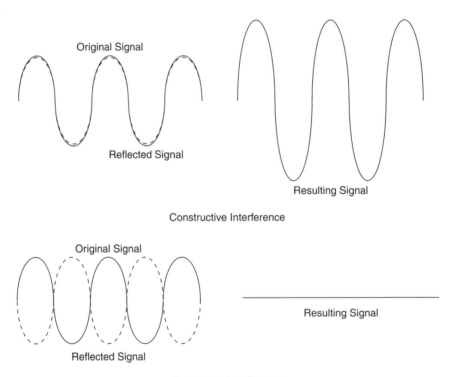

Figure 1-29   A multipath schematic.

A tag placed in one of the protruded regions will read, but if this tag moves slightly so that it is inside the surrounding dead region, the tag cannot be read (which might lead to nonintuitive tag-reading behavior). For example, when placed a certain distance away from a reader, a tag does not read, but when moved slightly in one direction, it can be read by the reader; if this tag is then moved slightly in another direction, however, it cannot be read! The read behavior of a tag near a protruded region is thus unreliable. Therefore, when you place an antenna to cover a read area, it is important that you *not* depend on these protruded regions to maximize the read distance. The best strategy is to stay inside the main ellipsoid-shaped region even if it means sacrificing the read range by a few feet—better safe than sorry.

It is extremely important to determine the antenna footprint; the antenna footprint determines where a tag can or cannot be read. The manufacturer might provide the antenna footprint as part of the antenna's specifications. However, you should use such information as a guideline only, because the actual footprint will most likely vary depending on the operating environment. You can use well-defined techniques such as *signal analysis* to map an antenna footprint. In signal analysis, the signal from the tag is measured, using equipment such as a *spectrum analyzer* and/or a *network analyzer*, under various conditions (for example, in free space, different tag orientations, and on conductive materials or absorptive materials). By analyzing these signal strengths, you can precisely determine the antenna footprint.

Antenna polarization, another important concept of reader antenna design, is discussed in the following section.

1.2.3.2 Antenna Polarization

As discussed previously, an antenna emits electromagnetic waves into its surroundings. The direction of oscillation of these electromagnetic waves is called the *polarization* of the antenna. What does this mean to tag readability? A great deal! The readability of a tag, together with its reading distance and reading robustness, greatly depends on the antenna polarization and the angle at which the tag is presented to the reader.

The main antenna types in UHF, based on polarization, are

- Linear polarized
- Circular polarized

The following subsections discuss these two types of antennas.

1.2.3.2.1 Linear Polarized Antenna

In this antenna type, the RF waves emanate in a linear patter from the antenna. These waves have only one energy field. Figure 1-30 shows the resulting wave pattern emanating from a linear polarized antenna.

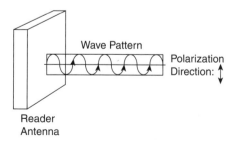

Figure 1-30 Wave pattern from a linear polarized antenna.

A linear polarized antenna has a narrower radiation beam with a longer read range compared to a circular polarized antenna. In addition, a narrower radiation beam helps a linear polarized antenna to read tags within a longer, narrow but well-defined read region (compared to a circular polarized antenna), instead of reading tags randomly from its surroundings. However, a linear polarized antenna is sensitive to tag orientation with respect to its polarization direction. These types of antenna are therefore useful in applications where the tag orientation is fixed and predictable. Figure 1-31 shows how a tag should be oriented with respect to a linear antenna for its proper reading in case of backscatter communication.

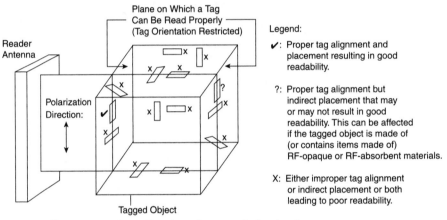

Figure 1-31   Proper tag orientation for a linear polarized antenna.

1.2.3.2.2   Circular Polarized Antenna

RF waves radiate from a circular polarized antenna in a circular pattern. These waves have two constituting energy fields that are equal in amplitude and magnitude, but have a phase difference of 90°. Therefore, when a wave of an energy field is at its highest value, the wave of the other field is at its lowest. Figure 1-32 shows the resulting wave pattern emanating from a circular polarized antenna.

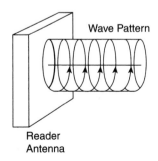

Figure 1-32   Wave pattern from a circular polarized antenna.

Because of the nature of polarization, a circular polarized antenna is largely unaffected by tag orientation. Therefore, this type of antenna proves ideal for applications where the tag orientation is unpredictable. A circular polarized antenna has a wider radiation beam and hence reads tags in a wider area compared to a linear polarized antenna. This antenna is preferred for an RFID system that uses high UHF or microwave frequencies in an operating environment where there is a high degree of RF reflectance (due to presence of metals and so forth). Figure 1-33 shows how a tag should be oriented with respect to a circular antenna for its proper reading in case of backscatter communication.

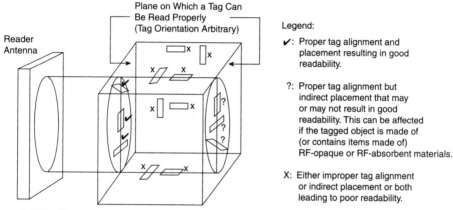

Figure 1-33   Proper tag orientation for a circular polarized antenna.

Figure 1-34 shows the circular and linear polarized antenna patterns.

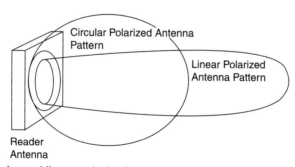

Figure 1-34   Circular and linear polarized antenna patterns.

Often, a *patch* antenna is used for making UHF antennas, as described in the following subsection.

1.2.3.2.3   Patch Antenna

A patch antenna, also called a *microstrip* or *planar antenna*, in its basic form consists of a rectangular metal foil or a plate mounted on a substrate such as Teflon. The other side of the substrate is coated with a metallic substance. A microstrip connected to the rectangular metal foil supplies power to the antenna (see Figure 1-35). The power supply type can be varied to make a patch antenna circular or linear polarized.

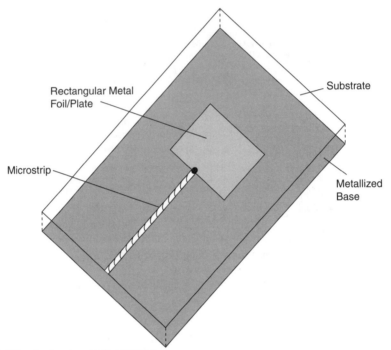

Figure 1-35   An basic patch antenna.

1.2.3.3   Antenna Power

An antenna emits power measured in either *effective radiated power* (ERP) units in Europe or in *equivalent isotropic radiated power* (EIRP) units in the United States. ERP and EIRP are not the same but are related by the relation EIRP = 1.64 ERP. The maximum possible value of antenna power is limited by national and international (for example, FCC in the United States) regulations. To use an antenna with higher power than the allowable limit, you must obtain explicit permission from the appropriate regulatory body. You can always reduce antenna power, however, by placing a small device called an *attenuator* in the transmission line (for example, between an antenna connector and the reader port). As a result, the antenna's signal strength is reduced, and the antenna's read range is diminished. Attenuation proves very useful in situations where the read zone needs to be constrained as a part of system requirements so that tags are only read inside but not outside this region. The ability of an attenuator to reduce the antenna strength varies depending on the attenuator.

1.2.4 Controller

A *controller* is an intermediary agent that allows an external entity to communicate with and control a reader's behavior together with the annunciators and actuators associated with this reader. A controller is the *only* component of an RFID system (or a reader, depending on point of view) through which reader communications are possible; no other medium or entity provides this ability. As mentioned previously, a controller for a reader can be embedded inside the reader or can be a separate component by itself. An analogy is in order. A controller to a reader is what a printer driver is for a computer printer. To print a document from a computer to a printer, the computer must have the appropriate printer driver software installed. Similarly, to retrieve tag data stored on a reader, a computer must use a controller—it cannot communicate to the reader in any other way.

A controller also provides (or uses, depending on viewpoint) a communication interface for the external entities to interact with it (as described previously in the section about readers).

1.2.5 Sensor, Annunciator, and Actuator

A reader does not have to be turned all the time; it can be started (and stopped) automatically if needed. A sensor can be attached with a reader for this purpose. This sensor can then be used to turn on/off the reader based on some external event detected by this sensor. A sensor can thus be used to provide some kind of input trigger to a reader.

An *annunciator* is an electronic signal or indicator. Examples of annunciators include audible alarms, strobes, light stacks, and so on. A light stack consists of a vertical arrangement of different-colored indicators and is useful for displaying various statuses of different system attributes. For example, the red indicator might mean invalid or bad tag data in the read zone, green might indicate a valid tag read, and amber might signal network connection between the reader and the controller is down. Figures 1-36 shows an example light stack.

An *actuator* is a mechanical device for controlling or moving objects. Examples of actuators include a *programmable logic controller* (PLC), robot arm, mechanical arm for an access gate, and so on. A PLC is one of the most versatile actuators, and PLCs are widely used in manufacturing plants. PLCs enable a variety of actions to be performed (such as monitoring and controlling a product packaging line, or applying a predetermined amount of torque to nuts in a mechanical assembly [for example, an automobile]).

Annunciators and actuators can thus be used to provide some kind of local output from an RFID system, such as audio-visual alarms in case of a read failure, opening an access gate for a successful read, and so forth.

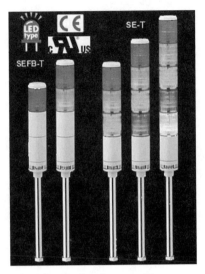

Figure 1-36   LED Signal Tower from Patlite Corporation.

Reprinted with permission from Patlite Corporation

1.2.6   Host and Software System

The *host and software system* is an all-encompassing term for the hardware and software component that is separate from the RFID hardware (that is, reader, tag, and antenna); the system is composed of the following four main components:

- Edge interface/system
- Middleware
- Enterprise back-end interface
- Enterprise back end

In a nontrivial RFID system, all these components are present to some degree. The following sections discuss these components.

1.2.6.1   Edge Interface/System

This component integrates the entire host and software system with the RFID hardware (which consists of the reader, tag, and antenna). This integration is accomplished by establishing communication with and control of the central nervous system of RFID hardware: the readers. Therefore, this component's main task is to get data from the readers, control the readers' behavior and use the readers to activate the associated external actuators and annunciators.

This component is logically and physically closest to the RFID hardware and can be considered to be at the edge when viewed from the host and software system perspective. Therefore, this is also the right place for this component to activate external actuators and annunciators without any need to go through the reader. This placement proves very useful because then the choice and control capabilities of annunciators and actuators are not limited by the reader support, but can be extended as and when needed by customizing the *edge system*.

The edge system is also the perfect place to hide the nitty-gritty details of interaction with a specific reader (through its controller) from a particular manufacturer. Therefore, this component also provides an abstraction layer for any type of readers needed by the RFID system. This abstraction layer is very desirable because then the rest of the host and software system can use this abstraction to interact with any supported readers, present and future, without any need to change itself.

This component can be viewed as a kind of a super controller that can be used to interact with any supported reader controller in the RFID system.

Moreover, this component can do several other tasks that are beyond the responsibilities of a simple controller, such as the following:

- Filter out duplicate reads from different readers

- Allow setting of event-based triggers that can automatically activate an annunciator or an actuator

- Provide intelligent functions such as aggregating and selectively sending out tag data to host and software system

- Remote reader management

- Remote management of itself

As apparent from the preceding discussion, this component may actually be hosted on specialized hardware as an embedded system. The rest of the host and software system can then interact with this embedded system over a wired or a wireless network. This component can be implemented using a standard such as *Open Services Gateway initiative* (OSGi), which defines a standard for dynamic delivery of software services to network devices (see Chapter 10). In a very simple case of a trivial, possibly throwaway, pilot, this component might be completely absent.

1.2.6.2   Middleware

The *middleware* component can be broadly defined as everything that lies between the edge interface and the enterprise back-end interface. This component can be viewed as the central nervous system of the RFID system from the software perspective (RFID readers can be considered the same from an RFID hardware perspective) in that it provides core functionality of the system, including the following:

- Data sharing both inside and outside of an enterprise

- Efficient management of massive data produced by RFID system

- Provide generic components that can be used as building blocks for implementing the business specific filtering and aggregation logic

- Open standard based so that it is compatible with a wide range of other software systems
- Enable loose coupling between the edge interface and the enterprise back-end interface (and thus any change in the former will minimally affect the latter)

In the extreme case of a trivial, possibly throwaway, pilot, this component might be completely absent.

This is the most complex and important component of the host and software system. As a result, a principle part of the implementation effort will be spent on implementing this component. Therefore, when implementing an RFID system, it is always preferable to procure this component as an off-the-shelf system from RFID software and services vendors. You can then customize it to meet the application requirements.

1.2.6.3   Enterprise Back-End Interface

The *enterprise back-end interface* component is used to integrate the middleware component with the enterprise back-end component. This is the place for implementing business process integration. Which processes need to be integrated with the RFID system will determine the amount of effort needed to implement this component. This effort can be substantial if business process changes are involved or comprehensive.

Because the middleware is a generic component, some customization is almost always needed to trigger transactions and transfer data between it and the enterprise back end. It is not uncommon to find enterprise-scale integration interfaces natively built in to the enterprise scale systems, such as ERP and WMS, that are available from large third-party software vendors.

1.2.6.4   Enterprise Back End

The enterprise back-end component encompasses the complete suite of applications and IT systems of an enterprise. This is thus the data repository and the business processes engine for the entire enterprise. In an RFID system context, this component provides the directory data for the tagged objects to the middleware component.

Note that in general, integration with a handful of applications or systems is necessary to achieve a satisfactory integration with the enterprise back end and hence the business processes. This is, of course, assuming that this component is well architected and implemented.

This component generally involves minimum effort from the implementation perspective of an RFID system because this is already built and functional. However, in some cases (for example, proprietary system elements), some effort might be necessary to actually modify or enhance this component to make it compatible with the RFID system that is being built.

1.2.7   Communication Infrastructure

This component provides connectivity and enables security and systems management functionalities for different components of an RFID system, and is therefore an integral part of the system. It includes the wired and wireless network, and serial connections between readers, controllers, and computers. The wireless network type can range from a *personal area network* (PAN, provided by Bluetooth), to a *local area network* (LAN, offered by 802.11x technology), to a *wide area network* (WAN, provided by 2.5G/3G technologies). Satellite communication networks, for example, using geosynchronous L-band satellites are also becoming an increasing reality for RFID systems that need to work in a very wide geographical area where existence of a pervasive reader infrastructure is not guaranteed.

It is now time to pause for a moment and learn about the basic concepts of an RFID system.

1.2.8   Basic Concepts

This section discusses the following terms that are frequently used in reference to an RFID system:

- Frequency
- Tag collision
- Reader collision
- Tag readability
- Read robustness

Frequency is the *most important* attribute of an RFID system. It has already been discussed in detail in the beginning of this chapter.

The remaining terms are discussed in detail now in the following subsections.

1.2.8.1   Tag Collision

Contrary to popular belief, a reader can only communicate with one tag at a time. When more than one tag attempts to communicate with the reader at the same time, a *tag collision* is said to occur. In this case, in response to the reader's query, multiple tags reflect back their signals at the same time to the reader, confusing it. A reader then needs to communicate with the conflicting tags using what is called a *singulation protocol*. The algorithm that is used to mediate tag collisions is called an *anti-collision algorithm*. Currently, the following two types of anti-collision algorithms are most widely used:

- ALOHA for HF
- Tree Walking for UHF

Using one of these anti-collision algorithms, a reader can identify several tags in its read zone in a very short period of time. Thus, it appears that this reader is communicating with these tags almost simultaneously.

1.2.8.2   Reader Collision

When the read zone (or read window) of two or more readers overlap, the signal from one reader can interfere with the signal from another. This phenomenon is called *reader collision*. This situation can arise if the antennas of these two readers are installed in such a manner that it gives rise to *destructive interference* (antenna footprint). As a result, RF energy from one of the antennas of a reader "cancels out" the RF energy from one of the antennas of the other reader. To avoid this problem, position the reader antennas so that the antenna of one reader does not directly face the antenna of another reader. If the direct facing of these antennas is unavoidable, separate them a sufficient distance so that their read zones do not overlap. You can use proper attenuators to attune the antenna power to achieve this. In addition, two antennas of the same reader can generally overlap without creating a reader collision, because the power to the antennas is physically transferred by the reader in such a manner that only one antenna is active at a time. As a result, there is no chance of two or more antennas of this reader emitting signals at the same time. You can also use another technique, called *time division multiple access* (TDMA), to avoid reader collision. In this scheme, the readers are instructed to read at different times rather than all reading at the same time. As a result, the antenna of only one reader is active at a time. The problem with this approach is that a tag can be read more than one time by different readers in the overlapping read zone. Therefore, some intelligent filtering mechanism must be implemented by the controller or the edge system/interface to filter out the duplicate tag reads.

1.2.8.3   Tag Readability

Tag readability of an RFID system for a particular operating environment can be defined as the capability of the system to read a specific tag data successfully. Tag readability depends on a number of factors (see Chapter 9, "Designing and Implementing an RFID Solution"). From a simple perspective, an RFID system needs to read a tag successfully just once to provide good tag readability. To make this guarantee, however, the system should be designed so that it can read a single tag several times, so that even if a tag read fails several times there's a good chance that one of the reads will succeed. In other words, an RFID system should have good read for robustness. This is the topic of the next section.

1.2.8.4   Read Robustness

Read robustness (also called *read redundancy*) is the number of times a particular tag can be read successfully when inside a read zone. As noted in the previous section, an RFID system has to be designed such that it has good read robustness for the tags. The speed of a tagged object can negatively impact the read robustness as the amount of time spent by the tag in the read zone decreases with an increase in its speed. This results in a decrease of read robustness for this tag. The number of tags present at one time in the read zone also can hamper read robustness because the number of tags that can be read by a reader per unit time is limited.

1.2.9   Characterization of an RFID System

An RFID system can be characterized in three different ways using the following attributes:

- Operating frequency
- Read range
- Physical coupling method

These criteria are interrelated. The first two criteria are most frequently used in practice. All three characterizations are discussed next.

1.2.9.1 Characterization Based on Operating Frequency

Operating frequency is the most important attribute of an RFID system. It is the frequency at which the reader transmits its signal. It is closely associated with the typical reading distance attribute. In most cases, the frequency of an RFID system is determined by its typical reading distance requirement. Frequency has already been described earlier in this chapter.

1.2.9.2 Characterization Based on Read Range

Read range of an RFID system is defined as the reading distance between the tag and the reader. Using this criterion, an RFID system can be divided into the following three types:

- Close coupled
- Remote coupled
- Long range

The following subsections describe these types.

1.2.9.2.1 Close-Coupled System

The read range of the RFID systems belonging to this class is less than 1 cm. The LF and HF RFID systems belong to this category.

1.2.9.2.2 Remote-Coupled System

The RFID systems belonging to this class have a read range of 1 cm to 100 cm. Again, this category contains LF and HF RFID systems.

1.2.9.2.3 Long-Range System

RFID systems having a read range of more than 100 cm belong to this class. RFID systems operating in the UHF and microwave frequency range belong to this group.

1.2.9.3 Characterization Based on Physical Coupling Method

Physical coupling refers to the method used for coupling the tag and the antenna (that is, the mechanism by which energy is transferred to the tag from the antenna). Based on this criterion, three different types of RFID systems are possible:

- Magnetic
- Electric
- Electromagnetic

These following subsections discuss these different types.

1.2.9.3.1   Magnetic-Coupled System

These types of RFID systems are also known as *inductive-coupled systems* or *inductive-radio systems*. The LF and HF RFID systems belong to this category.

1.2.9.3.2   Electric-Coupled System

These types of RFID systems are also known as *capacitive-coupled systems*. The LF and HF RFID systems belong to this category.

1.2.9.3.3   Electromagnetic-Coupled System

The majority of RFID systems belonging to this class are also called *backscatter systems*. RFID systems operating in the UHF and microwave frequency range belong to this group.

1.3   Conclusion

This chapter provided an in-depth discussion of the RFID basics. Admittedly, it covered a plethora of information that might be difficult to assimilate in a single reading. However, a good knowledge of this material is crucial to understanding the technology and applying it to solve real-world challenges. The material in this chapter is used throughout this book to develop related concepts. The reader is advised to revisit this chapter several times until confident of a good grasp of the fundamentals. If you want to dive deeper into the theoretical aspects of RFID technology, consult the *RFID Handbook*, Second Edition, by Klaus Finkenzeller (John Wiley & Sons, 2003).

CHAPTER 2

Advantages of the Technology

This chapter discusses the advantages of RFID technology in detail. Each particular advantage is examined from different perspectives to provide you with an in-depth understanding of the technology.

2.1 Advantages of RFID

The advantages of RFID can be broadly classified into the following two types:

- **Current.** These advantages are immediately realizable with the technology products that exist today.

- **Future.** These advantages are either available in some form today or will be available as improved features in the future as the technology matures.

These are not official terminologies, but are used for the sake of convenience and to aid in better understanding of a benefit. The following list covers both of these advantage types, and the rest of this chapter describes how much benefit is available today versus how much will be available in the future:

- **Contactless.** An RFID tag can be read without any physical contact between the tag and the reader.

- **Writable data.** The data of a read-write (RW) RFID tag can be rewritten a large number of times.

- **Absence of line of sight.** A line of sight is generally not required for an RFID reader to read an RFID tag.

- **Variety of read ranges.** An RFID tag can have a read ranges as small as few inches to as large as more than 100 feet.

- **Wide data-capacity range.** An RFID tag can store from a few bytes of data to virtually any amount of data.

- **Support for multiple tag reads.** It is possible to use an RFID reader to automatically read several RFID tags in its read zone within a short period of time.

- **Rugged.** RFID tags can sustain rough operational environment conditions to a fair extent.

- **Perform smart tasks.** Besides being a carrier and transmitter of data, an RFID tag can be designed to perform other duties (for example, measuring its surrounding conditions, such as temperature and pressure).

The following, although often touted as a benefit of RFID, is *not* considered an advantage:

- **Extreme read accuracy.** RFID is 100 percent accurate.

The following sections discuss the previously listed advantages in detail.

2.1.1   Contactless

An RFID tag does *not* need to establish physical contact with the reader to transmit its data, which proves advantageous from the following perspectives:

- **No wear and tear.** Absence of physical contact means there is no wear and tear on the readers as well as on the tags for reading and writing data.

- **No slowing down of operations.** Existing operations do not have to slow down to bear the extra overhead of bringing a reader physically into contact with a tag. Establishing such a physical contact can sometimes prove impossible. In a scenario in which tagged cases of items are moving at a rapid speed on a conveyer belt, there is a high chance that a reader will fail to maintain a physical contact with such a moving box, resulting in a missed tag read. As a result, had RFID been contact-based, it could not have been applied satisfactorily in a large number of business applications (such as supply-chain applications and so on).

- **Automatic reading of several tags in a short period of time.** Had RFID been contact-based, the number of tags read by a reader would have been limited by the number of tags it could touch at a particular time. To increase this number, the reader's physical dimensions need to be increased, resulting in a higher-cost, clumsy reader.

In summary, RFID instantly offers several benefits just by being contactless. In addition, this is clearly a current advantage of the technology.

2.1.2   Writable Data

RW RFID tags that are currently available can be rewritten from 10,000 times to 100,000 times or more! Although the use of these types of tags is currently limited compared to *write once, read many* (WORM) tags, you can use these tags in custom applications where, for example, time-stamped data about the tagged object might need to be stored on the tag locally. This guarantees that the data will be available even in absence of a back-end connection. In addition, if a tag (that is currently attached to an object) can be recycled, the original tag data can be overwritten with new data, thus allowing the tag to be reused. Although writable tags might seem like an advantage, they are not widely used today because of the following reasons:

- **Business justification of tag recycling.** Virtually all business cases that involve tag recycling impact business operations. For example, the following must be factored in: how tags are going to be collected from the existing objects, when they are going to be collected, how these are going to be re-introduced to the operations, additional resources and overhead required, and so on. Unless the tag is active or semi-active and is expensive, in most situations, generally, tag recycling does not make business sense.

- **Security issue.** How can tags safeguard accidental and malicious overwriting of data by valid and rogue readers when in use? If the application is used outside an enterprise in an uncontrolled environment, the security implications multiply many times. Even if such a tag is used within the four walls of an enterprise, the issue of security remains. To satisfactorily address this issue, additional hardware, setup, and processes might be necessary; this, in turn, can result in high implementation costs that might prove unjustifiable. Currently, it seems as if RW tags will continue to be used within the specific secure bounds of an enterprise.

- **Necessity of dynamic writes.** If most of the RW tag applications are going to be used mainly inside the four walls of an enterprise, there is a high degree of probability of the presence of a network and the ability to access the back-end system through this network. Therefore, using the unique tag ID, the back end can store the data without any need to write this data on the tag itself. Also, process changes can be made to handle exceptional conditions when the network is down—for example, generally critical manufacturing facilities have two modes of operation, one automatic and one manual so that if the automatic mode of operation fails, the operators can switch to the manual mode without stopping production lines.

- **Slower operating speed.** A tag write is often slower than a tag read operation. Therefore, an application that does tag rewrites has a good possibility of being slower compared to an application that does tag reads only.

These issues might seem daunting to the reader. However, it is certainly possible that some RFID applications exist for which using RW tags makes good business as well as technical sense. An example of such an application is monitoring the production quality control of a bottling operation for a medical drug. First, RW RFID tags are attached to empty bottles, which are then washed in hot water and sanitizing solutions, dried, and subsequently go through a series of steps before the drug is placed in these bottles and sealed. It is assumed that the tags are sturdy enough to withstand the various processing steps. At each processing step, the parameters of the process—such as temperature, humidity, and so on—are written to the tags. When the sealed bottles roll off the assembly line, their associated tag data is automatically read by quality control systems. This way, any processing step that fell short of the minimum requirements can be discovered, and the overall quality of the bottling process can be quantized.

This is a current advantage of RFID that also offers future advantages in terms of better data security and improved technology.

2.1.3   Absence of Line of Sight

The absence of line of sight is probably the most distinguishing feature of RFID. An RFID reader can read a tag *through* obstructing materials that are RF-lucent for the frequency used. For example, if a tag is placed inside a cardboard box, a reader operating in UHF can read this tag even if this box is sealed on all the sides! This capacity proves useful for inspecting the content of a container *without* opening it. This feature of RFID has privacy rights infringement implications, however. If a person is carrying some tagged items in a bag, an RFID reader can (*potentially*) read the tagged item data without this person's consent. If this person's personal information is associated with the tagged item data (at the point of sale by the merchant, for example), it might be possible to access this information (using a suitable application) without the person's consent or knowledge, which might constitute a privacy rights infringement. To prevent this, a reader should not read these tags after sale is completed unless explicitly needed or authorized by the buyer. There are multiple ways to achieve this objective (see Chapter 5, "Privacy Concerns"). Note that in some situations, a line of sight is needed to help configure the tag read distance, reader energy, and reader antenna to counter the environmental impact. These situations involve UHF tags and the presence of a large amount of RF-reflecting materials, such as metal, in the operating environment giving rise to multipath (see Chapter 1, "Technology Overview"). For example, consider a machinery tool production line where virtually everything is made of metal. A large amount of RF energy from the readers installed in this environment gets reflected from the objects in the environment. In this case, to achieve a good read accuracy, a tag and a reader must be placed so that there is no obstacle between them.

This is a current advantage of RFID. It is possible that future improvements in the technology can bypass some of the hurdles faced by the presence of RF-opaque materials between the reader and the tag. Therefore, this is a future benefit, too.

2.1.4   Variety of Read Ranges

A *low-frequency* (LF) passive RFID tag generally has a read distance of a few inches; for a passive *high-frequency* (HF) tag, this distance is about 3 feet. The reading distance of an *ultra-high-frequency* (UHF) passive tag is about 30 feet. A UHF (for example, 433 MHz) active tag can be read at a distance of 300 feet and an active tag in the gigahertz range can have a reading distance of more than 100 feet. These reading distances are usually realized under ideal conditions. Therefore, the actual tag-reading distance of a real-world RFID system can be substantially less than these numbers. For example, the reading distance of 13.56 MHz tags in general do not exceed a few inches. This wide array of reading distances makes it possible to apply RFID to a wide variety of applications. Whereas the LF read distance passive tags are ideally suited for security, personnel identification, and electronic payments, to name a few, you can use HF passive tags for smart-shelf applications; passive UHF for supply-chain applications, tracking, and many other types of applications; and, finally, you can use passive tags in the microwave ranges for anti-counterfeiting. You can use active and semi-active tags in these frequency ranges for tracking, electronics toll payment, and almost limitless other possibilities. As you can understand, RFID has virtually an unlimited spectrum of current and possible applications.

Today, the tags for every frequency type are commercially available. In addition, the location of an active or a passive tag can be associated with a reader that reads this tag. Therefore, if a reader installed at a certain dock door of a warehouse reads a tag in its read zone, the location of this tag can be assumed to be this dock door at the time of reading. This location information can then be made available through a private or public (for example, Internet) network over a wide geographical area. As a result, the tag can be tracked thousands of miles away from its actual location. Future improvements of the technology will have limited impact on this aspect because the entire range of reading distances is currently available using direct (that is, a reader) and indirect (that is, a network) means. Hence, this feature is a current advantage of RFID.

2.1.5   Wide Data-Capacity Range

A typical passive tag can contain a few bits to hundreds of bits for data storage. Some passive tags can carry even more data. For example, the ME-Y2000 series (also known as *coil-on chip*) passive, RW miniature tag from Maxell (see Figure 2-1) operating in the 13.56 MHz range can carry up to 4 K bytes of data within its 2.5 mm × 2.5 mm space.

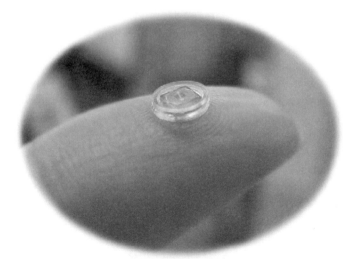

Figure 2-1   HF tag from Maxell Corporation of America.

Reprinted with permission from Maxell Corporation of America

An active tag has no theoretical data-storage limit because the physical dimensions and capabilities of an active tag are not limited, provided this tag is deployable.

There are two approaches to use an RFID tag for an application. The first one stores only a unique identification number on the tag, analogous to a "license plate" of an automobile that uniquely identifies the tagged item; the second one stores both a unique identification number and data related to the tagged object. A large number of unique identifiers can be generated with a relatively small number of bits. For example, using 96 bits, a total of 80,000 trillion trillion unique identifiers can be generated (see Chapter 10, "Standards")! So, a relatively small number of bits are sufficient to tag virtually any type of object in the world. However, some applications might choose to store additional data on a tag locally. The advantage of storing this data locally is that no access to a networked database is required to retrieve the object data using its unique identifier as a key, an advantage that proves useful if the tagged object is going to be moved around in areas where the presence of network access to an object database is either not available or undesirable. Even when such a network connection is available, the associated application is such that it must not be impacted by a network outage or delay. Therefore, one of the benefits of storing data locally on the tags is that the resulting application can be made largely independent of a back-end system. However, such a scheme has drawbacks compared to a "license plate" type of approach. First, data security needs to be addressed so that tag data can neither be accidentally overwritten by a valid reader nor by a rogue reader intentionally. The transmission time necessary for a high data capacity tag to transmit all its data bits correctly to a reader can be several times more compared to just transmitting the unique identifier. In addition, an increase in data transmission leads to an increase in error rate of transmission. A high memory capacity tag will be more expensive than the tags that can store only a unique identifier. Therefore, just because it is available, using a high memory capacity tag in an application does *not* seem like a good idea unless the application specifically demands it (especially true for applications that have a hard time limit to perform a specific task). An active tag, however, can use a large data-storage capacity to support its custom tasks. A small amount of which, most probably containing the results of these tasks, might end up getting transmitted by this tag (which is perfectly acceptable because this data is dynamic and can only be determined by the tag itself by scanning its environment).

This is a future benefit. Most of the passive tags available on the market today are constrained in memory size. These tags are used in "license plate" types of applications, and therefore they prove quite adequate for the task at hand. More high memory capacity tags will become available in the future.

2.1.6   Support for Multiple Tag Reads

Support for multiple tag reads ranks as one of the most important benefits of RFID. Using what is called an *anti-collision algorithm*, an RFID reader can automatically read several tags in its read zone in a short period of time. Generally, using this scheme a reader can uniquely identify a few to several tags per second depending on the tag and the application. This benefit allows the data from a collection of tagged objects, whether stationary or in motion (within the reader limits), to be read by a reader, thus obviating any need to read one tag at a time. Consider, for example, one of the classic tasks of a financial institution: counting a stack of currency notes to determine its total count and value. Assuming these notes have proper RFID tags, the data from these currency tags can be read using an RFID reader, which can then be used to determine the total count and the value of the notes in aggregate in a very short period of time, automatically. This method is much more efficient compared to the traditional counting techniques. Now consider another classic example: loading a truck with cases of merchandise at a shipping dock and receiving it at a receiving dock. Currently, for these types of applications, either the boxes are not inventoried at all during shipping time (they are, however, inventoried most of the time at the receiving dock) or they are inventoried using bar codes (which is manual and time-consuming). As a result, business might lose a considerable amount of inventory annually due to shrinkage or incur a high recurrent overhead in the cost of labor. If RFID tags can be applied to the boxes before they are shipped, a stationary reader placed near a loading truck can read all the boxes, automatically, when these boxes are being loaded into this truck. Thus, the business can have an accurate list of items being shipped to a distributor or a retailer. In addition, significant labor costs were saved by eliminating manual scanning of the labels, which would have been unavoidable if a technology such as bar code had been used instead. The data collected from these tags can be checked against the actual order to verify whether a box should be loaded into this truck (thus reducing the number of invalid shipments). As you can understand, this particular RFID advantage can speed up and streamline existing business operations considerably.

Contrary to popular belief, a reader can communicate with only *one* tag in its read zone at a time. If more than one tag attempts to communicate to the reader at the same time, a *tag collision* occurs. A reader has to resolve this collision to properly identify all the tags in its read zone. Therefore, a reader imposes rules on communication so that only one tag can communicate to the reader at a time, during which period the other tags must remain silent. This is what constitutes an anti-collision algorithm (see Chapter 5, "Privacy Concerns"). Note that there is a difference between reading a tag's data in response to an anti-collision command versus reading a tag's data completely. In the former case, only certain data bits of a tag are read; whereas in the latter, the complete set of data bits of a particular tag are read. In addition, there is a theoretical as well as practical limitation on how many tags can be identified by a reader within a certain period of time.

This is a current benefit, but it is possible that future improvements in RFID reader technology might substantially increase the number of tags that can be identified per second (within the theoretical and practical limits).

2.1.7 Rugged

A passive RFID tag has few moving parts and can therefore be made to withstand environmental conditions such as heat, humidity, corrosive chemicals, mechanical vibration, and shock (to a fair degree). For example, some passive tags can survive temperatures ranging from –40°F to 400°F (–40° C to 204°C). Generally, these tags are made depending on the operating environment of a specific application. Today, *no* single tag can withstand *all* these environmental conditions. An active and semi-active tag that has on-board electronics with a battery is generally more suscepti-ble to damage compared to a passive tag. A tag's ruggedness almost always increases its price.

This is a current benefit because tags with a variety of resistance to operating environments are available. However, plenty of room exists for improvement, and as the tag technology improves, it is expected that more tags will be available that can better resist harsh environments than their present-day counterparts. Therefore, this can also be called a future benefit.

2.1.8   Perform Smart Tasks

The on-board electronics and power supply of an active tag can be used to perform specialized tasks such as monitoring its surrounding environment (for example, detecting motion). The tag can then use this data to dynamically determine other parameters and transmit this data to an available reader. For example, suppose that an active tag is attached to a high-value item for theft detection. Assume that this active tag has a built-in motion sensor. If someone attempts to move the asset, the tag senses movement and starts broadcasting this event into its surroundings. A reader can receive this information and forward the information to a theft-detection application, which in turn can sound an alarm to alert the personnel. It might seem that by just taking off the tag from the asset and then putting the tag back where it was (while taking the asset away) would fool the tag into thinking that the asset is stationary again. However, it is possible for such a tag to sense that it is no longer attached to the asset. The tag can then send another type of broadcast message to signify this event.

This aspect of RFID has the greatest potential for improvement as active tags with special-ized functionalities are becoming available. Hence, this can be called a future benefit.

2.1.9   Read Accuracy

In the media, the read accuracy of RFID is mentioned variously as "very accurate," "100 percent accurate," and so on, but no objective study shows how accurate RFID reads really are. It would definitely be desirable to back up such accuracy statements with hard data, because no technol-ogy can offer 100 percent read accuracy in every operating environment all the time. Factors on which RFID read accuracy depends include the following:

- **Tag type.** Which frequency tags are being used, the tag antenna design, and so on can have a bearing on the read accuracy of an RFID system.

- **Tagged object.** The composition of the object, how it is packed, the packing material, and so on play important roles in determining the readability and hence the read accu-racy. Also note that impact of this factor depends on the frequency of the RFID system used.

- **Operating environment.** Interference from existing mobile equipment, electrostatic discharge (ESD), the presence of metal and liquid bodies, among other factors, can pose a problem for read accuracy in the UHF and microwave frequencies.

- **Consistency.** Tag orientation and placement relative to the reader antennas can signifi-cantly impact read accuracy.

Another issue with RFID is what are called *phantom reads* or *false reads*. In this situation, a random but seemingly valid tag data is recorded by the reader for a brief period of time. After this time, the tag data can no longer be read by the reader! The problem arises when a reader receives incorrect data from a tag, which might happen for various reasons (such as a poorly constructed error-correcting protocol). Phantom reads are "bugs" in the supplier system. Incorrect installations might also give rise to this phenomenon. In general, phantom reads are not an issue. However, this shows that the objective determination of RFID accuracy is not easy, that it depends on several factors. It is possible for the accuracy rates of two identical RFID systems used in different environments to differ. It might not always be possible to increase the read accuracy and degree of automation of highly automated systems that are in existence today.

This is a current benefit because several applications generally do show sufficient accuracy to meet business requirements. However, the read accuracy of RFID has good potential to improve as improved tags, readers, and antennas become available in the future. Therefore, this can also be called a future benefit.

2.2 Conclusion

RFID offers several benefits, a good many of which you can realize today to a substantial degree with the existing products. Other benefits are also somewhat realizable, and it is expected that improvements in the technology will steadily bring these benefits to a mature level. Undeniably, however, such a collection of unique RFID benefits makes is already a potential enabler of a wide variety of applications. Some of the benefits have privacy rights infringement implications that might present issues regarding RFID use in some situations. Even with these issues, however, RFID will likely be the preferred technology in other areas. RFID technology is currently undergoing rapid changes that are expected to provide continuously improving products, steadily bringing the true potential of the technology to the user.

Limitations of the Technology

RFID is not without its limitations. This chapter discusses the limitations of RFID. Note that some of the limitations that exist today have good potential to be overcome as technology improvements take place. Therefore, you are advised to look at these limitations as opportunities for creative solutions.

3.1 Limitations of RFID

The current limitations of RFID include the following:

- **Poor performance with RF-opaque and RF-absorbent objects.** This is a frequency-dependent behavior. The current technology does not work well with these materials and, in some cases, fail completely.

- **Impacted by environmental factors.** Surrounding conditions can greatly impact RFID solutions.

- **Limitation on actual tag reads.** A practical limit applies as to how many tags can be read within a particular time.

- **Impacted by hardware interference.** An RFID solution can be negatively impacted if the hardware setup (for example, antenna placement and orientation) is not done properly.

- **Limited penetrating power of the RF energy.** Although RFID does not need line of sight, there is a limit as to how deep the RF energy can reach, even through RF-lucent objects.

- **Immature technology.** Although it is good news that the RFID technology is undergoing rapid changes, those changes can spell inconvenience for the unwary.

The remainder of this chapter discusses these limitations in detail.

3.1.1   Poor Performance with RF-Opaque and RF-Absorbent Objects

If high UHF and microwave frequencies are used, and if the tagged object is made of RF-opaque material such as metal, some type of RF-absorbent material such as water, or if the object is packaged inside such RF-opaque material, an RFID reader might partially or completely fail to read the tag data. Custom tags are available that alleviate some of the read problems for particular types of RF-opaque and RF-absorbent materials. In addition, packaging can present problems if made of RF-opaque materials such as metal foils.

It is expected that improvement in the tag technology will overcome several of the current problems associated with RF-opaque/RF-absorbent objects.

3.1.2   Impacted by Environmental Factors

If the operations environment has large amounts of metal, liquids, and so on, those might affect the read accuracy of the tags, depending on the frequency. The reflection of reader antenna signals on RF-opaque objects causes what is known as *multipath* (see Chapter 1, "Technology Overview"). It is a safe bet in these types of environments to provide a direct line of sight to the tags from a reader. Although the tag reading distance, reader energy, and reader antenna configuration are the major parameters that need to be configured in these cases to counter the environmental impact, a line of sight helps to achieve this configuration. In some cases, however, this might not be possible (for example, in an operating environment where there is high human traffic). A human body contains a large amount of water, which is RF-absorbent at high UHF and microwave frequencies. Therefore, when a person is in between a tag and a reader, there is a good possibility that this reader cannot read the tag before this person moves away. So, serious degradation of system performance might result. In addition, the existence of almost any type of wireless network within the operating environment can interfere with the reader operation. Electric motors and motor controllers can also act as a source of noise that can impact a reader's performance. Some older *wireless LANs* (WLANs) in the 900 MHz range can interfere with the readers. This problem mostly exists in older facilities that have not upgraded their WLAN equipment.

This issue most probably will remain for some time. Some of the present interference issues might remain unresolved.

3.1.3   Limitations on Actual Tag Reads

The number of tags that a reader can identify uniquely per unit time (for example, per second) is limited. For example, today, a reader on average can uniquely identify a few to several tags per second. To achieve this number, this reader has to read tags' responses several hundred times a second. Why? Because the reader has to employ some kind of anti-collision algorithm to identify these tags; to identify a single tag, a reader has to walk down the range of possible values (see Chapter 5, "Privacy Concerns"). Therefore, several readings of tag responses are required before a reader can uniquely determine tag data. A limit applies as to how many such reads a reader can perform within a unit time, which, in turn, dictates a limit on the number of unique tags that can be identified within this same time period.

Improvement in the reader technology will undoubtedly increase the number of tags that can be uniquely identified per unit time, but there will always be an ultimate limit on this number that no reader will be able to exceed.

3.1.4 Impacted by Hardware Interference

RFID readers can exhibit *reader collision* (see Chapter 1) if improperly installed. A reader collision happens when the coverage areas of two readers overlap and the signal of one reader interferes with the other in this common coverage area. This issue must be taken into account when an RFID installation plan is worked out. Otherwise, degradation of system performance might take place.

This issue can be somewhat solved today by using what is known as *time division multiple access* (TDMA). This technique instructs each reader to read at different times rather than both reading at the same time. As a result, two readers interfere with one another no longer. However, a tag in the overlapping area of these two readers might be read twice. Therefore, the RFID application must have an intelligent filtering mechanism to eliminate duplicate tag reads. As RFID technology improves, new solutions to this issue might become available.

3.1.5 Limited Penetrating Power of the RF Energy

The penetrating power of RF energy finally depends on the transmitter power of the reader and duty cycle, which are regulated in several countries around the world (see Chapter 1). For example, a reader might fail to read some cases on a pallet if they are stacked too deep, even if these cases are all made of RF-lucent material for the frequency used. How many such cases can be put on a pallet for proper reading? You can only determine the answer to this question by experimenting with actual boxes stacked on an actual pallet in the actual operating environment using actual RFID hardware. This number will also vary from country to country, depending on the restriction of reader power and duty cycle. Therefore, the answer needs to be determined experimentally; it is very difficult, if not impossible, to determine it theoretically.

Unfortunately, this limitation will remain as long as international restrictions on reader power and duty cycle remain. Therefore, if you need to deploy a solution in multiple countries, you must seriously consider this issue.

3.1.6 Immature Technology

Immature technology is a practical issue facing RFID technology today. An RFID solution can only be as good as the hardware currently available from the vendors. The vendors are doing their best to develop improved products, but maturity might not be available for some time to come. For example, it is not uncommon for passive UHF tags currently used in supply-chain operations to get damaged when subjected to existing handling techniques, with the defective tag rates shooting up as high as 20 percent or more.

The same tag types (for example, passive 915 MHz for metal) from different vendors might perform differently. It is possible that a tag needed for satisfactory read accuracy for a specific application is not currently available, despite intensive research being conducted both at the theoretical (for example, antenna design) and manufacturing (for example, material used, processing techniques) levels to build tags of different kinds. Building a custom tag for an application can be very expensive (typically in the range of $100,000).

Readers have come a long way in the past two years, gradually transitioning from a simple interrogator to a well-defined network device with built-in intelligence to support several functions needed by an RFID application. Some of these functions are filtering, caching recent tag reads, input from external sensors, output for activating sensors and actuators, and so on. Similarly, antenna technology is making antennas smaller and cheaper. However, a side effect of these improvements is that new RFID hardware is coming out at a very fast rate, which might urge businesses to upgrade their equipment at the same rate! Such rapid upgrades are generally not necessary because the products are backward compatible in most cases. Business has to implement realistic strategies so that its current investment in RFID hardware is not wasted as new products come out.

The issue of immaturity/maturity will continue to be part of RFID technology in the near future. Stabilization of the technology in terms of products and globally acceptable standards will eradicate this issue, but a prediction of the timeline for such stabilization is difficult to make.

3.2   Conclusion

You might now feel a little worried after reviewing the limitations of RFID in this chapter. The reality is not so bleak! Although the technology does have several drawbacks, it is not common to encounter a total showstopper because of any of these issues. In most cases, either the issues can be bypassed, or a good RFID solution can be implemented if proper attention is paid to some basic aspects. Chapter 9, "Designing and Implementing an RFID Solution," provides more details on optimizing this technology.

CHAPTER 4

Application Areas

The potential application of RFID technology is limited only to one's imagination. Although a popular belief holds that RFID is best suited to supply-chain management or *consumer packaged goods* (CPG) industries, the range of current RFID applications goes far beyond these areas. In fact, a variety of established RFID application types have already been deployed successfully in real-world environments. An *application type* consists of several different application members that share the same characteristics of the application type. These application types are denoted by the term *prevalent* in this book to emphasize that these are the most commonly applied areas of RFID today. Note, however, that this is not an official term, but is used for the sake of convenience.

The full potential offered by RFID does not stop at the prevalent application types. RFID is an emerging technology, and as such, tremendous potential lies ahead to apply it to areas that can utilize the benefits of the technology. Currently, some of these areas are in the prototyping or planning stage, some have just started to be explored, and others are not yet getting sufficient attention from the industry or from vendors. In short, these application types need to mature to a point of validation (both technological and relating to business processes) before being rolled out into production. This book calls these application types *emerging*, to highlight the fact that most of these application types have yet to become prevalent application types. Again, the term *emerging* is not an official term, but is used here for convenience.

Whereas a whole application type can be emerging, some members of a particular prevalent application type can be emerging, too (item tracking and tracing, for example, and inventory monitoring and control, discussed later in this chapter). When RFID begins to be used in these types of applications, a whole new suite of yet undiscovered applications might unfold. An application at a prototype stage today might very well be rolled out as a business application in the near future. Therefore, the application types and members belonging to prevalent and emerging types will vary over time based on technological advances and the willingness of the business community to apply the technology to solve new problems.

This chapter discusses examples of both *prevalent* and *emerging* applications, as of this writing, to give you an idea of the application potential of the technology. This book uses the terms prevalent and emerging for both application types and members belonging to application types. You will understand the distinction between these two entities from the context. Note that this chapter does not attempt to list every RFID application that exists today or that might exist in the future. Instead, this chapter covers some of the most important applications that exist today and that might be possible in the future, with the hope that you can identify similar applications and associate them with the ones discussed here.

4.1 Prevalent Application Types

Currently, some of the most prominent RFID applications are as follows:

- Item tracking and tracing
- Inventory monitoring and control
- Asset monitoring and management
- Anti-theft
- Electronic payment
- Access control
- Anti-tampering

The application types listed here do not appear in any particular order. The order in which an application appears in this list does *not* reflect, for example, its importance or the degree of its applicability in the RFID context. In addition, these listed applications are not mutually exclusive in terms of characteristics and benefits. Item tracking and tracing applications can have inventory monitoring and control, anti-theft, and asset management characteristics and benefits. The following sections discuss each of these application types in detail. For each type, at least one concrete member application is examined. For each such example, the benefits and caveats are discussed. An implementation note accompanying each such member application provides a brief implementation-specific detail. Finally, some real-world deployments of these solutions are also provided optionally.

4.1.1 Item Tracking and Tracing

The *item tracking and tracing* application class type is characterized by the following:

- Attaching a tag containing a unique identifier on an item to be tracked
- Reading this tag identifier at specific locations while the item moves

The tag identifier, when associated with the reading time and the location information, can provide near real-time information about the whereabouts of this item at a particular point in time.

You can use a list of such location information to track the object's movement during its life cycle. You can also capture additional information, such as which personnel moved the object from one location to another. This information can prove useful, for example, to determine the personnel responsible for shrinkage, if any. You can also associate various actions with this tracking activity, such as triggering an alarm if an object is not spotted at a location at a certain time. Two technical solutions currently apply to item tracking:

- **Satellite tracking.** Satellite tracking can identify the location of the tag whenever queried.

    However, an RFID tag that can communicate directly via satellite communication has yet to appear on the market. In addition, this is quite distinct from the capabilities offered by RFID. It is possible for an RFID tag to indirectly "communicate" to the base station via a reader via satellite communication or some type of wired/wireless network. Note that in this context, commercially available products today can perform two-way communication directly via satellite communication. For example, the T2000 from TransCore Corporation is a rugged, ultra-compact satellite communication device that has a built-in antenna (see Figure 4-1). This device can perform secure two-way communication using geosynchronous L-band satellites. To reiterate, however, this is not an (active) RFID tag.

Figure 4-1   GlobalWave T2000 Data Communication Terminal.

Reprinted with permission from TransCore

- **Limited-range active/passive tagging.** Limited-range active/passive tagging requires the tag to move through a *choke* point to be scanned.

If a customer chooses the second option and there are an insufficient number of choke points, the solution might not provide the needed performance. In contrast, the solution might not be economically feasible if there are too many choke points.

Member Applications

Some of the most important applications that belong to this class type are as follows:

- **Supply-chain management.** This is discussed in the next subsection.

- **Tracking of hazardous materials.** This is discussed in the next subsection.

- **Airline baggage tracking.** RFID tags embedded in airline baggage tags can be used to provide an effective tracking solution. Such an RFID tag has sufficient storage to carry baggage handling and routing data so that this data is available locally, bypassing any need to access a baggage database. RFID tags can be read, unlike barcodes, in virtually any orientation (irrespective of overlaps with other baggage), resulting in faster and more accurate scanning as compared to bar codes. The *International Air Transport Association* (IATA) has yet to adopt an industry standard to replace bar coded luggage tags with RFID and automatic handling of passenger baggage. In industry testing (British Airways in 1999 and Delta Airlines in 2003) of the technology, RFID tag labels resulted in accuracy rates in the 95 percent to 99 percent range, whereas bar codes could only offer accuracy rates in the range of 80 percent to 85 percent (approximately). This application has not yet been widely deployed commercially. Airline baggage tracking is an example of an emerging application member belonging to this prevalent application type (that is, item tracking and tracing).

The following subsections discuss the first two examples of this application type.

4.1.1.1   Supply-Chain Management

An item can be tracked in the supply chain from where it is produced to the point where it is consumed or recycled. A plastic container of motor oil, for example, can be tagged at the point of production with a tag that contains a unique identification number.

NOTE

The popular "can of cola" example is deliberately avoided here. UHF is generally the most preferred RFID frequency used in supply-chain operations today. A can of cola is made of metal, which is RF-opaque and contains a potable liquid, which is RF-absorbent in the UHF frequency range of operation. This combination makes it very difficult, if not impossible, to tag a can of cola so that it can be properly read every time in the various stages of its life cycle (assuming UHF is used).

A container of motor oil, on the other hand, is made of plastic, which is RF-lucent, and contains motor oil, which (again) is RF-lucent in UHF. Therefore, a container of motor oil can be realistically tagged with good read results.

That container can then be tracked by reading the tag data at the following points in the supply chain:

1. At the manufacturer's shipping dock, the container is loaded onto a truck that will leave the manufacturing plant.

2. The container arrives at the receiving dock of the distribution center.

3. At the distribution center's shipping dock, the container is loaded onto a truck that will leave the distribution center.

4. The container arrives at the retailer.

5. The customer buys this container at the sales counter of the retailer.

6. The empty container arrives at a recycling center.

Figure 4-2 shows these example read points.

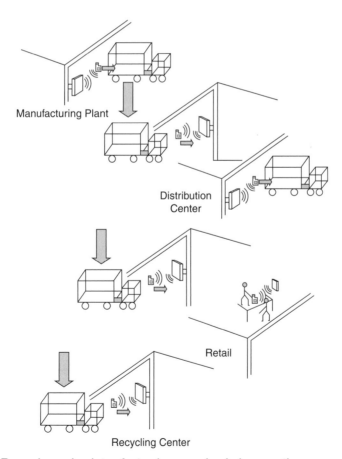

Figure 4-2 Example read points of a tag in a supply-chain operation.

The tag data could also be read at other points in the supply chain. For example, the tag data could be read when storing this container at a particular location inside the warehouse in the distribution center, or at a particular storage area inside the retailer. Such a reading enables personnel of the particular location to locate this container inside its four walls. A retail shelf reader can also detect the physical presence of an item placed on the shelf. A back-end system can use this information to determine whether this shelf needs to be replenished with new containers of motor oil. A reader cannot make this decision by itself; it can only report its tag list to the host application. The host application can then look for a specific item type (that should be on the shelf) based on its tag ID. If a matching item type is not found or if the tag list is empty, the application might determine that the shelf needs to be restocked. Figure 4-3 shows the simple logic involved in determining an out-of-stock situation.

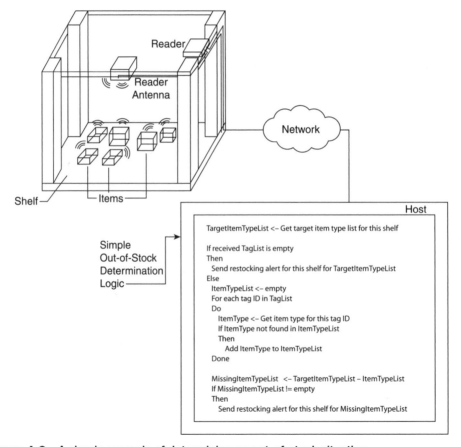

Figure 4-3   A simple example of determining an out-of-stock situation.

It is widely believed that if the tag price drops below 5¢, unit-level tagging applications become viable. However, several obstacles currently block the realization of this goal, such as the right business models, manufacturing issues, privacy concerns, and implementation complexity.

Benefits

As you can understand from the preceding discussion, RFID offers *item-level visibility* in supply-chain management. RFID can supply businesses with accurate and real-time information, which can result in the following benefits:

- **Reduce shrinkage for the manufacturer.** Because an item can be tracked through the entire supply chain, and the information gathered can include the personnel moving the item, the responsible parties and the point(s) of occurrence can easily be traced in case of shrinkage.

- **Enable the retailer to better understand a product's sale potential.** Customers who buy a certain item can be tracked, and businesses can use this data to target buyers for special promotions. (Note that a bar code solution can also offer similar benefits.)

- **Enable better inventory management for the retailer.** The retailer, better able to understand a product's sale potential, can stock up or stock down a particular item, thus maximizing the sales potential with optimum inventory.

- **Enable better asset monitoring and utilization.** The ability to accurately locate an item and its associated information (which can be both static and dynamic), enables businesses to better monitor and utilize this item.

- **Reduce out-of-stocks for the retailer.** When a customer removes an item from the shelf, the back-end system(s) detects the absence of the item. Businesses can use this information to determine whether the shelf is empty and needs to be replenished.

Caveats

- Tagging and tracking, down to an individual item level, generally represents the most challenging task in the RFID context because of virtually infinite variations in the tagged object material type, packaging, environmental conditions, and so on. Generally, a pallet is the easiest item to tag, followed by a case, because a pallet has limited variations in shape, size, and composition material as compared to that of an item. These factors vary more for a case, but are still limited as compared to that of an item. However, depending on the read requirements and environmental conditions, the degree of complexity involved in reading pallet, case, and item tags can vary (see Chapter 9, "Designing and Implementing an RFID Solution").

- A significant privacy concern is raised by the tagging of individual items that a consumer can buy or use. The concern is that a consumer may be tracked via the tags on the items he or she buys (see Chapter 5, "Privacy Concerns"). Businesses are currently wary of adopting item-level tagging before the privacy issues are settled.

Implementation Notes

A passive tag is almost always used for this type of application. The passive tags in the UHF frequency range (868–870 MHz and 902–928 MHz) offer the best tradeoff between reading distance and price and, therefore, are extensively used for this purpose. Tracking a large number of items presents some of its own unique challenges, such as the generation of a tremendous amount of raw data that must be processed and moved through the enterprise network. The most widely accepted solution for this type of application is provided by the EPCglobal specification (see Chapter 10, "Standards"). This specification provides a cheap, open, and interoperable standard for readers and tags. It also provides an architecture that tackles the challenges presented by this application type. In short, EPCglobal is a powerful and flexible solution that has the potential to be accepted as a worldwide standard.

Installed Base

Large manufacturers, retailers, and government entities around the world are employing this type of application to track items at case and pallet levels. Individual item-level tagging seems to be taking a back seat because its cost and privacy impacts remain unclear. (*Unsettled* is probably a better word.) The current trend is clearly in favor of case- and pallet-level tagging, because privacy issues are fewer, the cost of implementation and implementation complexity are low (as compared to item-level tagging), and the return on investment is quick. Note, however, that in some stores, customers tend to buy cases of items (besides individual items). In these situations, case-level tagging might also pose a privacy concern for some customers.

4.1.1.2   Hazardous Materials Tracking

Chemical processing plants handle many different kinds of chemicals on a daily basis. Chemicals arrive from different suppliers and are then consumed or processed in the plant. The plant ships the products manufactured from these chemicals to distributors and customers. The used chemicals are recycled. Some of the chemicals can be hazardous and, therefore, special care must be taken when handling these. With regard to a hazardous chemical container received from a supplier, it is very desirable that some of the following critical information always be available:

- What is the chemical type, its constituents, and other properties such as concentration and so forth (that is, the *Material Safety Data Sheet* [MSDS] information)?
- When did the container arrive at the shipping dock?
- Who ordered it?
- When and where was it last spotted inside the plant?

- Is it in transit inside the plant? If so, what is the source and what is the destination? Has it arrived at the destination within the estimated time?

- Has it exited the plant (for example, been shipped to the original supplier for recycling, or been shipped to a distributor or a customer)?

Benefits

- **Public safety.** Proper tracking can prevent hazardous materials from being handled inappropriately. A small quantity of a hazardous material, if mishandled or having fallen into the wrong hands, can cause substantial financial and emotional damage, both for a plant and those exposed. At this time of heightened national security, every chemical plant that handles hazardous materials should implement strict access and handling controls over their chemicals.

- **Less environmental pollution.** Proper recycling and decontamination of hazardous materials and their containers can help prevent environmental pollution.

Caveats

- It is very desirable to track a hazardous material at the item level, in addition to the pallet and case levels.

- Tagging and tracking hazardous materials in metal cylinders and plastic containers in liquid form might prove difficult (see Chapter 9).

Implementation Notes

Generally, passive tags in the 13.56 MHz and 915 MHz ranges are used. At the moment, there is no pressing need to require the exchange of item tracking information beyond the four walls of the plant. Therefore, these types of solutions tend to be *closed-loop systems*, although this might change in the future. Generally, the most important information about the chemical is also stored on the tag so that it can be read locally together with its ID so that this crucial information about the chemical is always available even if the network connection to the back end (which contains data about this chemical) goes down. Specialized metal tags are available that can tag a metal container. Privacy issues are less of a concern here, even though the items are being tracked individually because of the very nature of these items.

Installed Base

IBM has rolled out a unified RFID chemical container tracking system in its manufacturing facilities in Burlington, Vermont; Fishkill, New York; and Bromont, Quebec, Canada. Chemicals in plastic drums and metal cylinders are tracked from the time they are received until the associated container is either decontaminated or sent back to the supplier. The system also helps to determine whether the right chemical is being used in the processing equipment.

In November 2004, NASA Dryden Flight Research Center successfully completed a 90-day test of a real-time hazardous chemical tracking network called ChemSecure that uses passive RFID tags. ChemSecure is being developed to provide hazardous chemical tracking at five NASA facilities in Southern California at Edwards Air Force Base.

4.1.2   Inventory Monitoring and Control

The *inventory monitoring and control* application class type is characterized by the following:

- Attaching a tag containing a unique identifier on an inventory item to be monitored
- Detecting the presence or absence of this item in the inventory by attempting to read the tag data on a periodic basis

When an item is placed in inventory, the tag data is read by a stationary reader, which then transmits the tag data and its location (based on this reader's location) to the back-end inventory system. The back-end inventory system registers the item in the inventory database. While physically in inventory, the reader (which has this tag in its read zone) periodically transmits all read tags in its read zone to the inventory system. If the back end does not receive a registered item's tag data corresponding to this reader, the back end assumes that the item is absent from the inventory. If this item's absence results in an out-of-stock situation for this item type, the inventory system can take the following actions:

- Automatically notify personnel and other associated systems
- Post an order for this item to its supplier(s)

Refer to Figure 4-3 to see an example of the logic involved in determining out-of-stock situations.

Member Applications

This application class type can be considered a variation of the track and trace. However, a distinguishing feature of this application type is that it is always tracked in the context of an inventory. That is, an item is either in the inventory or is not. Some example member applications belonging to this type are as follows:

- **Smart shelves.** This is described in the next subsection.
- **Parts inventory management for airline and automobile manufacturing plants.** Large airlines can track about half a million parts, for example, and might have as much as $1 billion worth of parts in inventory. The current inventory process is overwhelmingly manual, resulting in a high degree of error (and hence a high cost of maintaining inventory). An RFID tag used for such a purpose needs to have a large amount of memory. (For example, it is not unusual to have tags with 10 K of memory.) This extra

memory is needed to store custom data of a part, such as repair history and part identification. RFID tags that operate in the 13.56 MHz frequency range prove most suitable for this purpose because they can be used in metal environments and this frequency has worldwide approval. However, these types of tags have a low read/write range (less than 1 meter). This is an example of an emerging application member belonging to this prevalent application type.

The following subsection examines smart shelves, an interesting application.

4.1.2.1   Smart Shelves

Today, stocking shelves is generally a manual process, one that is often less than optimal. In a smart-shelf application, a tagged item is placed on a store shelf. A single reader or multiple stationary readers mounted on or near the shelf monitor the presence of the tag (and hence the item). When a consumer removes this item from the shelf, the reader(s) can no longer read the tag. Therefore, the tag lists reported by the reader(s) to the inventory system no longer contain this tag. The inventory system then assumes that the item has been removed from the shelf. The inventory system can automatically perform several actions, such as notifying the store personnel to replace more items of the same type to avoid an out-of-stock situation. Refer to Figure 4-3 to see a simple example implementation.

Benefits

- Reduce out-of-stocks for the retailer.
- Enable the retailer to better understand a product's sale potential.
- Enable better inventory management for the retailer.
- If an item is misplaced, the back-end system(s) can notify store personnel where this (misplaced) item is and where to place it correctly.
- Some level of support for anti-theft. If an unusual number of items are suddenly missing from a shelf, it *might* be possible that a theft has occurred.

Caveats

- A smart-shelf application tracks inventory at an individual item level; pallets and cases do not count. As a result, this type of application represents the most challenging task in the RFID context (see Chapter 9).
- The tags are still too expensive to be used for tagging individual items that are not high-valued not to mention the cost of other hardware like the readers and antennas. The cost of the tag has to fall to less than one penny for its widespread use to tag low-value items. The current goal and challenge for manufacturers is to produce a 5¢ tag, which may take another 10 to 15 years or more.

- The implementation is far from simple, as discussed next.

- Substantial privacy concerns abound regarding the tagging of individual items that a consumer can buy or use. What is the guarantee that retailers and manufacturers will not monitor their products after they have been sold to customers (see Chapter 5)?

Implementation Notes

Generally, passive tags in the 13.56 MHz range are used to tag individual items on a shelf. Configuring the readers and antennas on the shelves can prove tricky. The number of items that can be read needs to be maximized independent of the items' orientations on a shelf, which generally requires several antennas per shelf, which in turn can introduce overlapping read zones and can cause interference. This situation is not desirable because the antennas must read items on the shelf on which they are installed and should not have overlapping read zones with other antennas on a different shelf. If the items are packed too densely on a shelf, the stationary readers might not be able to read all the items on the shelf, which might lead to inventory issues.

In addition, customers might pick up an item and put it back on the same shelf but with the tag attached to the item oriented suboptimally to the reader antenna(s). Such misplacement might also occur when store personnel load up the shelf in such a way that some of the items are misaligned with the reader antenna(s). If the reader(s) cannot properly read a tag, the inventory system will incorrectly assume that the item has been removed. A customer might also pick up an item and then place it on another shelf, from which the reader associated with the original shelf cannot read this tag. In such a scenario, if the inventory system is not intelligent enough to look at the misplaced items on other shelves, it will incorrectly assume that this item has been removed.

Shelves are generally made of metal, which detrimentally affects tag reads. Short-range readers operating at 13.56 MHz are used to alleviate this problem, which introduces two other problems: First, multiple readers might be necessary to cover a long shelf, which increases the hardware cost; second, duplicate reading of the same item by different readers might occur. Therefore, if the RFID middleware is not intelligent enough to correctly filter out the duplicate readings, the inventory system might experience inconsistencies. In short, the implementation of smart shelf is not straightforward, and such implementation can prove expensive. Therefore, *currently*, it might not make business sense for a retailer to implement this application on a large scale. Even though the smart-shelf applications might not be rolled out in the near future, this does not mean that vendors and businesses should avoid investigating the potential for such applications in their relative environments.

4.1.3   Asset Monitoring and Management

The *asset monitoring and management* application class type is characterized by the following:

- Attaching a tag that contains a unique identifier on an asset item to be monitored

- Detecting the location and other properties and states of this item in real time by attempting to read the tag data on a periodic as well as on an on-demand basis

The basis of this class of application is the determination of the location of an item in real-time using RFID tags. The entire frequency range of RFID has asset-related products. You can use both passive and active tags for asset monitoring. In this context, note that an ANSI standard already exists. The ANS INCITS 371 standard developed by the *International Committee for Information Technology Standards* enables users to locate, manage, and optimize mobile assets throughout the supply chain (see Chapter 10). Generally, stationary readers read the asset tags when they pass through a certain facility. This data and the readers' location information are then transferred to the back end and fed into an asset-monitoring system. Both local and global/wide-area monitoring is possible. You can use satellite communication networks to link RFID systems for global asset monitoring; the major vendors that offer asset-monitoring solutions have either bought or partnered with at least one satellite communications company. Note that today, no such (active) tag exists that can perform satellite communication directly. However, a reader or a network of readers can be connected to a base station that, in turn, can use satellite communication. You can also use wireless 802.11x networks for local monitoring.

Member Applications

This application class type has a large overlap with item tracking and tracing. Indeed, an item to be tracked can be viewed as an asset that can be monitored. However, one distinguishing aspect of this type of application is collection of asset properties in real time, together with its unique ID, to aid in management of this asset. One example is collection of vehicle diagnostic data together with the vehicle's unique ID to manage the life cycle of a fleet better. Some important examples of this application type are as follows:

- **Fleet monitoring and management.** This is discussed in detail in the next subsection.
- **Animal tracking.** Today, use of RFID is becoming common to track livestock. A tag attached to an animal can be used to monitor its health, movement, and so on. Animal tracking can also be used to track wildlife and fish to monitor their characteristics (such as migration and breeding patterns). ISO 11784 /11785, the international standard for radio frequency identification of animals based on 134.2 KHz technology, is the prominent standard for animal tracking. Some criticize this standard being susceptible to duplicate identification numbers that can be introduced by different manufacturers due to lack of proper enforcement of the identification numbers by the standard bodies. The *International Committee for Animal Recording* (ICAR) is a Paris-based international body that is responsible for worldwide standardization of animal recording and productivity evaluation. ICAR, in agreement with the ISO, has been developing compliance procedures for testing and validation of RFID systems with this ISO standard.

4.1.3.1   Fleet Monitoring and Management

In this type of application, RFID tags are mounted on transportation items such as power units, trailers, containers, dollies, and vehicles. These tags contain pertinent data about the item by which it can be identified and managed. Readers, both stationary and mobile, are placed at locations through which these tagged items move (for example, access controlled gates, fuel pumps, dock doors, and maintenance areas). These readers automatically read the data from the tags and transmit it to distributed or centralized data centers as well as an asset-management system. This system can then allow or deny a vehicle access to a gate, fuel, maintenance facilities, and so on. Thus, using the data from the tagged items and vehicles, an asset-management system can locate, control, and manage resources to optimize utilization on a continuous, real-time basis. The data captured from the tagged items is fast and accurate, resulting in elimination of manual entry methods, which, in turn, reduces wait times in lanes and dwell times for drivers and equipment.

Benefits

- **Better use of assets.** The ability to locate, control, and use an asset when needed allows optimum use of an asset in a fleet.
- **Improved operations.** Accurate and automatic data capture coupled with intelligent control leads to better security of controlled areas, provides proactive vehicle maintenance, and enhances fleet life.
- **Improved communication.** Real-time, accurate fleet data provides better communication to customers, management, and operation personnel.

Caveats

- **High initial investment might be needed in hardware and infrastructure.** Cost increases with the fleet size and the number of data capture points. In addition, for geographically dispersed operations, wide-area wireless communications such as satellite communication might be needed, thus increasing the infrastructure cost. Vendors generally supply a fleet-management system. Otherwise, the cost of custom implementation of such a system can be expensive.

Implementation Notes

Semi-active read-only, and read-write tags with specialized on-board electronics (for example, to indicate the status of a data transaction), are generally used. Most importantly, such a tag can be integrated with a vehicle's on-board sensors to relay critical vehicle information such as fuel level, oil pressure, and temperature to a reader. The fleet-management system uses this data to determine proactive maintenance on vehicles, resulting in a longer fleet life.

Installed Base

Fleet-management systems using RFID have been deployed by the Maryland Transit Administration (Automatic Vehicle Location System).

4.1.4 Anti-Theft

RFID can provide an effective deterrent against theft. A solution of this type is characterized by the following:

- Attaching a tag to an item to be monitored for theft
- Reading the tag ID at the vulnerable points (for example, at exit points, during starting of the ignition of an automobile, and so on)
- Alternative or additional features such as the ability to remove an attached tag from the item only after the *correct* payment has been made, the ability to detect movement of the attached item and reporting it to a nearby reader, and so on

You can use both passive and active tags for this purpose. For a high-value item such as a laptop, an active tag with an built-in motion detector can be attached to the item. Whenever this laptop is moved, the built-in motion detector in the tag can sense the motion and transmit this information to its surroundings. An appropriate reader can receive and relay this information to a back-end system. The back end can then initiate various actions. For example, it can either lock the exit(s) through which the item can be taken out of the building, it can trigger an alarm, or it can initiate a video recording of the place where the item is currently located. If a passive tag is used, its tag ID can be read at an exit point, or the absence of this tag can be detected by the back-end system using stationary readers (attached to the ceilings in the storage area, for instance). This, in turn, can trigger multiple actions by a back-end system. Note that RFID anti-theft solutions are currently not cheap. Therefore, the cost of implementing an anti-theft RFID solution needs to be carefully weighed against the benefits.

Retail is a very important area for anti-theft applications. According to the University of Florida National Retail Security Survey (2002), U.S. retailers lose an estimated $31.3 billion from inventory shrinkage.[1] Consumer theft accounts for 31.7 percent of this loss. In addition, the Center For Retail Research (based in the United Kingdom), estimates that shrinkage costs Western European retailers about €30 billion annually.[2] About 48 percent of these losses result from customer theft. These losses directly translate into lost revenue and a thinner profit margin for retailers already in a fiercely competitive marketplace. The use of RFID is presently gaining momentum in retail anti-theft applications. The application class type called EAS (*electronic article surveillance*), described in the following section, deserves a special mention in this

---

[1] National Retail Security Survey, Final Report. University of Florida, 2002. Richard C. Hollinger, Ph.D., director, and Jason L. Davis, graduate research associate.

[2] Key results of the European Retail Theft Barometer, 2004. J. Bamfield, Centre for Retail Research (Nottingham, UK).

context because its use is so widespread and well established today. EAS widely uses RF tags that cannot be called RFID tags per se. However, RFID tags can be used in conjunction with EAS tags to enhance the anti-theft capabilities of an EAS system.

Member Applications

The following are some example RFID anti-theft applications in use today:

- **Automotive anti-theft immobilization.** In this commercially deployed solution, an embedded reader located inside the car (for example, in the steering wheel) becomes activated when a driver turns the ignition key. This reader then attempts to read the valid unique code from a tag in its vicinity. Generally, the tags are small and can be embedded in an ignition key (for example, in the key head). If the reader detects a valid tag, the ignition starts. In general, passive 134.2 KHz LF tags are used. However, today's car thieves can breach these systems in various ways (for instance, by locating and overriding the embedded RFID unit, or by using a device that can imitate the code transmitted by such a key). To counter these methods, new-generation RFID anti-theft applications use a combination of active and passive tags that involve multiple authentication steps. The automobile cannot be started until all authentication steps have succeeded. Therefore, even if a potential car thief overrides one or a few of these steps, the other steps will prohibit the thief from starting the car and driving it away.

- **Retail anti-theft in combination with EAS.** This solution has also been commercially deployed. In this application type, a specialized RFID tag is attached to the item that already has an EAS RF tag. Generally, passive 13.56 MHz HF tags are used. Although this is an example of item-level tagging, the tags are removed at the *point of sale* (POS) and reused, thus bypassing any privacy concerns. Such an RFID tag, unlike the attached RF EAS tag, can report exactly what item has been shoplifted in case of a theft. Therefore, this application can also help the retailer in inventory management. The specialized features of the tag might include mechanisms to foil what is termed as *sweethearting*. In a sweethearting scenario, checkout personnel remove the tag without accepting any payment or accepting a sum that is less than the item's sale price. To prevent this, the attached RFID tag can be removed only when the correct payment has been received. The tag memory is also erased at the time of detachment so that it can be reused.

Now a brief discussion of EAS is in order.

4.1.4.1   Electronic Article Surveillance (EAS)

The application uses what are called 1-bit RF tags, or EAS tags, consisting of only 1 bit of storage. Thus, *no* unique item identification data is stored on the tag, and hence these tags cannot be used to identify an item uniquely. Therefore, these tags can be called RF tags, but *not* RFID tags. Initially, when a tag of this type is attached to an item to be monitored for theft, its bit value is set to 1. The bit value of this tag is set to 0 at the checkout counter when the customer pays for this item. When this tag is presented to a reader (generally, located at the exit points of a store), the tag notifies its presence by transmitting its bit value. A bit value of 1 signifies its presence, and a bit value of 0 signifies absence. When a reader determines the presence of such a tag in its read zone, it assumes that the associated item is being stolen. It then triggers visual and audible alarms to warn of a possible theft attempt. You can integrate EAS tags into item labels and price tags with no visible difference to the labels or price tags.

Benefits

- **Affordable solution.** The 1-bit tags are very cheap to produce and cost less than a penny.
- **Simple solution.** It involves minimum complexity as far as an RF solution is concerned.
- **Effective solution.** You can blend an EAS tag into merchandise physical characteristics in such a manner that it can be very difficult to detect it.
- **No privacy issues.** Because these tags do not contain any unique item identification, they cannot be used to track purchases bought by a customer.
- **Widely deployed solution.** The simplicity and cost of the solution are major drivers for its wide acceptance.

Caveats

- **Simple solution.** This solution can be easily defeated. A tag is generally hidden inside an item so that it cannot be easily detected. If a would-be thief can locate the tag, however, he can just strip the tag off the item and discard it, after which he can take the item out of the store without raising any alarms.

Implementation Notes

EAS tags are generally unaffected by pressure or magnets and are available in various and custom sizes to fit a particular product needs. The tags can generally be deactivated at a distance of 15 inches (38 centimeters approximately) and can be read as far away as 6 feet (about 1.8 meter). A reusable tag has a hard-to-defeat locking mechanism that is used to keep it attached to an item. With a detacher device, store personnel can easily detach these tags the POS. Disposable and reusable tags can be used together.

Installed Base

Almost every type of retail store, ranging from general merchandise stores to high-end electronics, use this solution to protect its merchandise from being stolen by shoplifters.

4.1.5   Electronic Payment

This *electronic payment* application class type is characterized by the following:

- A tag that contains a unique customer number
- Reading this customer tag data at the POS

At the time of transaction, the customer identification data on the tag is associated with the actual customer account number at the back end. This level of indirection protects customer account numbers in case the tag is missing or stolen. When a reader at the POS reads the customer identification data from the tag and the associated customer account number is located, the transaction then proceeds normally like any other regular transaction.

Member Applications

This application class type is gaining wide user acceptance, as indicated by the size of user bases of some of the applications belonging to this class (for example, Speedpass from Exxon-Mobil). This application class represents one of the massive-scale rollouts of an RFID solution, which is not so common even today. Some of the most important applications belonging to this class type are as follows:

- Speedpass
- Electronic toll payment

These two applications are probably the most well-known RFID electronic payment applications in use today and are discussed in the next subsections.

4.1.5.1   Speedpass

Speedpass, a very popular application from ExxonMobil, uses a "wand," which is a small cylinder shaped device that contains a tag, for electronic payment at participating Exxon and Mobil gas stations. A customer just points or waves this tag near a specially marked area on a pump or register. The application automatically initiates and completes the transaction without any need for the customer to punch in a PIN or sign a receipt.

Benefits

- **Fast, easy, and convenient.** All the customer has to do is point the tag appropriately; the rest is automatic. The tag can be carried in a key chain and is sturdy.
- **Cashless.** This small device obviates the need to carry cash because it is tied to the customer's account number, much like a debit card.
- **Secure.** Unlike a debit card, a customer's actual account number is never stored on the tag. In addition, the transaction type and the amount of money that can be spent per transaction are generally limited. Therefore, if this device is stolen, the maximum

amount of money that can be lost is limited. The issuer can invalidate a lost tag's data instantly, thus making the stolen tag useless, without changing the customer's actual account number.

- **High customer satisfaction.** More than 90 percent of customers report they are highly satisfied using the application.

Caveats

- **Limited acceptance.** The type of transactions that can be made using a tag are specialized and limited to participating Exxon and Mobil gas stations.

Implementation Notes

A "wand" is a passive, LF 134.2 KHz key-ring tag from Texas Instruments, Inc. It contains a 23 millimeter glass-encapsulated tag packaged inside plastic housing to withstand rough handling. It weighs about 4.5 grams. This tag is available in three versions, offering various levels of security. Figure 4-4 shows a Speedpass tag.

Installed Base

Started in 1997 by Mobil Oil Corporation, Speedpass is currently used at more than 8,900 locations in the United States and 1,600 locations in Canada, Singapore, and Japan. There are more than 6 million devices in the United States. There is no fee to enroll in the Speedpass program or to use Speedpass for purchases.

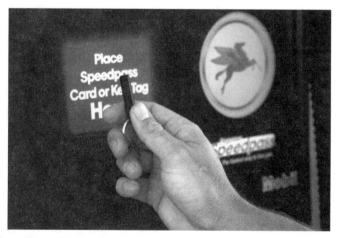

Figure 4-4   Speedpass tag.

Reprinted with permission from Texas Instruments

4.1.5.2   Electronic Toll Payment

Toll agencies in the United States and other countries use RFID to allow drivers to pay for tolls electronically at toll booths. A customer opens an account with a predefined amount of money with an agency that is responsible for toll collection. The customer then receives a tag with a unique ID. This tag is mounted, generally on the vehicle's windshield (see Figure 4-5), so that it can be read properly by readers at toll booths. When this customer drives through a toll booth that accepts electronic payment, the tag ID is read, the associated prepaid account is accessed, and the toll amount is subtracted from the account balance—automatically. The tag can display the account status by turning on its different-colored indicators. For example, green means toll paid, yellow means toll paid but account balance is low, and so on. A customer can fund his prepaid account with a credit card that automatically gets charged when the account balance is insufficient. Alternatively, a customer can also replenish his account online, via phone, or can mail a check.

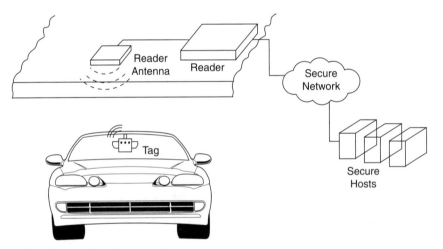

Figure 4-5   Electronic toll payment.

Benefits

- **Fast, easy, and convenient.** All the customer has to do is drive through the toll booth; the rest is automatic—no need to stop, carry exact change, or wait at a booth to get change from the operator.
- **Cashless.** This small device obviates the need to carry cash because it is loaded with a predefined amount of money. The device can be reloaded when the money it contains is used up.

- **Secure.** The tag contains a limited amount of cash. Therefore, if this device is stolen, the maximum amount of money that can be lost is limited. The issuer can invalidate a lost tag's data instantly, thus making the stolen tag useless. The overall system is also considerably secure, although the system back end was breached at least once (but was fixed very quickly). Personal "studies" on the Web advise how to bypass this system. However, you should *not* follow these schemes; other means are typically used to make sure this system is close to foolproof (for example, video recording of a vehicle's license plate even if the toll is paid, and police officers at standby to catch toll evaders).

Caveats

- **Limited acceptance.** Transaction types are limited to a specific toll agency's toll booths.

Implementation Notes

Generally, a semi-active tag is used, which besides containing the unique customer ID also contains specialized electronics for displaying account status, battery level, and so forth.

Installed Base

Some of the well-known electronic toll payment systems are SunPass in Florida; E-Z Pass in New Jersey, New York, New Hampshire, Maine, and Delaware; I-Pass in Illinois; Smart Tag in Virginia; CruiseCard in Georgia; FAST LANE system in Massachusetts; PIKEPASS in Oklahoma; and a combined electronic toll and traffic-management system in Houston.

4.1.6 Access Control

RFID has been successfully used in providing access control solutions. A solution of this type is characterized by the following:

- A tag that contains unique identification data and that is carried by the object or the person to gain access (for example, a tag placed on a vehicle windshield, embedded in an person's ID badge or under his skin)
- Reading the tag ID data at the access control points (with the ID then being forwarded to a security system that decides the actual access permission)

This application type is relatively mature compared to some other prevalent application types in terms of the RFID technology and systems that go with it. One of the characteristics of a mature technology is the existence of standards. The ISO 15693 (ISO SC17/WG8) vicinity cards standard is widely accepted by 13.56 MHz access control products.

Member Applications

Some well-known examples of applications belonging to this type are as follows:

- **Perimeter and building security systems.** This is discussed in the next subsection.
- **Parking access systems.** In this application, a passive RFID tag is attached to a vehicle (for example, on the windshield) that needs access to the parking system. When the driver pulls this vehicle up to the parking entrance, a reader reads the unique tag data on the tag and forwards it to an access system. This system, depending on the access permission associated with the tag data, either grants or denies access to the parking area. If access is granted, the entrance gates are opened to let the vehicle inside the parking area. Generally, 13.56 MHz passive tags are used for this application. Active and semi-active RFID tags are also used when long range and enhanced security are needed.

The first member application mentioned here is one of the most widely deployed. This is discussed in the next subsection.

4.1.6.1   Perimeter and Building Security System

This RFID application member is used for securing access to specific areas of a location (for example, a loading dock of a warehouse or a building). An example of the latter is the entry gate of a highly secured building (for example, an army headquarters), access to which, if compromised, could lead to danger of the personnel working in the building (besides the negative publicity that might result from such an incident). In July 2004, the Mexican government announced that it was implanting RFID tags under the skin of the employees of its $30 million anti-crime computer center in Mexico City to ensure secure access to this facility and to track an employee if he or she is kidnapped.[3]

Benefits

- **Flexible security control.** The permissions associated with a particular ID for a particular facility can be granted or revoked dynamically based on a central control system. Such an ID is first forwarded (via an RFID reader connected to a network) to this central security system. This system then uses a variety of factors, such as the number of access times to this facility by this ID and so forth, to decide whether permission should be granted. In addition, you can integrate this type of application with other RFID application types, such as anti-theft, to provide monitoring and tracking of persons who have accessed assets. For example, if the access control system determines that a person does

---

[3] The tracking benefit is misconceived because it confuses passive RFID tags with active RFID tags having GPS capability. Such tags do not currently exist, and these will most likely be too large to implant under the skin of a human.

not have the necessary permission, it can send a trigger to the anti-theft system to take action (for example, starting a video recording of this person and sending alerts to appropriate personnel and so on).

- **Fairly economical.** A new generation of cheaper ISO 15693 RFID tags have arrived on the market.

- **Relatively mature.** This application is well understood and has a wide variety of tag hardware available from vendors and systems from integrators.

- **Standard-based solution.** ISO 15693 is the de facto standard used for these types of tags.

Caveats

- **Can be bypassed.** An individual can tailgate an authorized person through an access point to bypass security.

- **Can be defeated.** The tag can be stolen and used to gain access provided the original tag owner is unaware of this theft (which would mean that the application has not been notified to deactivate this tag ID).

Implementation Notes

Generally, passive tags in the 128 KHz and 13.56 MHz frequency ranges compliant with the ISO 15693 standard are used for access control. The storage capacities of the tags can vary from 64 bits to 2 K bits, with read range of up to approximately 4.5 feet (1.5 meters). Note that passive tags in the UHF frequency ranges can also be used for this application member.

Installed Base

The 75th Academy Awards in Los Angeles used a passive 13.56 MHz RFID security system to provide access control to about 11,000 attendees. The U.S. Army prototyped this application in 2003.

4.1.7   Anti-Tampering

The *anti-tampering* application class type is characterized by the following:

- Attaching a tag that contains a unique identifier on an item to be monitored for anti-tampering. This tag is placed around the cap or lid of the container of this item, essentially forming an electronic seal.

- Detecting the occurrence of breaching the seal. Various methods can be used to determine the breach (for example, comparing the original position of the label with very high accuracy or using optical characteristics of the seal).

An RFID tag used for this application type acts as a tamperproof seal. This seal, besides identifying the sealed item uniquely, can also provide the evidence of tampering, if any. Moreover, this application type, when integrated with other application types such as access control, can also provide identification of the person(s) who might be involved in tampering with the seal. Although it might seem that this system could be defeated easily by carefully opening the seal and resealing it again, this is very difficult, and close to impossible in most cases. Both passive and active tags are used for members of this application type. This area is one of the busiest areas of research, and it is expected that sophisticated anti-tamper RFID tags will be available commercially in the future.

Member Applications

This application class type has gained the interest of the government, especially in the wake of heightened national security. The potential of this application type has also caught the attention of drug manufacturers. Currently, some of the application members belonging to this type are as follows:

- **Smart shipping containers and port security.** This is discussed in the next subsection.

- **Drug anti-tampering.** This is still in the prototyping stage and is considered an emerging application member. The tags used for this application member are passive tags in the 13.56 MHz frequency range. The tags are generally not impacted by x-rays or electronic methods of defeat. After a drug container is sealed, the tag can detect tampering if the tag position is altered, even by a very small amount. If the tag is carefully removed, it has to be put back extremely accurately in its original position to avoid detection by the tag. This accuracy is very difficult, if not impossible, to achieve in practice.

The first application member is currently experiencing strong backing from the government, and is discussed in the next subsection.

4.1.7.1   Smart Shipping Containers

This application member is used to secure containers shipped internationally. About 90 percent of the world's cargo moves by container. About 7 million cargo containers arrive and are offloaded at U.S. seaports each year. Less than 2 percent of these containers are opened by U.S. customs officials, which means this represents a potential area of concern for national security. In January 2002, U.S. Customs (now U.S. Customs and Border Protection) announced an initiative called the *Container Security Initiative* (CSI) to secure containers that could be a potential threat to global trade. The four basic elements of this initiative are as follows:[4]

---

[4]   "Extending the Nation's Zone of Security." Maggie Myers, Public Affairs Specialist. Office of Public Affairs, March 2004.

1. Using intelligence and automated information to identify and target containers that pose a potential terrorism risk

2. Prescreening containers as high-risk at the port of departure, before they arrive at U.S. ports

3 Using detection technology to quickly prescreen containers that pose a risk

4. Using "smarter," tamper-evident containers

The fourth element of the CSI is where RFID can play a major role. At present, 19 of the world's largest 20 ports have committed to join the CSI. These ports handle about two thirds of containers that enter the U.S. annually. CSI is also operational in Sweden and South Africa, with Malaysia and Sri Lanka having agreed to join the CSI effort. Currently, CSI signing discussions are underway for ports in South and Central America, Southeast Asia, Europe, and the Middle East.[5]

Closely related to CSI is the *Smart and Secure Tradelanes* (SST) initiative targeted at improving container security through a combination of technologies, including RFID, sensors, GPS, and so forth, integrated using the *Universal Data Appliance Protocol* (UDAP). The U.S. military has deployed the world's largest active RFID tag application, called *Total Asset Visibility* (TAV), which tracks and manages 270,000 military supply containers in 400 locations in 40 countries.

This application type can provide a good measure of security in an automatic manner over a wide geographical area. An RFID active tag is attached to the container and in some cases, besides having tamper-detection capabilities, can detect certain other conditions (biological weapons, narcotics, and explosives, for instance).

Benefits

- **Flexible security control.** These active RFID tags can function as sensors to detect the presence of explosives or radioactive emissions, among other things. In addition, this application member can be integrated with other application types, such as access control, to provide information about persons who have accessed this container.

- **Real-time notification of tampering.** If a shipping container has been tampered with, it might be of little use to discover this at a port. The tampered-with container might now contain extremely dangerous material, such as a biological or nuclear weapon, that can cause widespread destruction even when detected at this point in time. Potentially, RFID tags can be made to transmit via readers to a central station as soon as a tag detects tampering. The base station can then forward this information to a remote center using a wireless wide area network. This scheme does not need a costly setup (as compared to

---

[5] "Extending the Nation's Zone of Security." Maggie Myers, Public Affairs Specialist. Office of Public Affairs, March 2004.

setting up a network of RFID readers over a very large area to track an item) and can send tamper alerts almost anywhere in the world.

- **Privacy concerns are not an issue.** Although it might seem that tagging an item for tamper detection might intrude upon the privacy of the shipper(s), it is not applicable in this situation. According to international import and export regulations, authorities in charge may open any shipped container for inspection.

- **Real-time inventory.** This is one of the chief benefits, besides providing tamperproof capabilities. The readers inside a dock can receive data from these tags when the containers are in station, which results in automatic inventory.

- **Real-time location.** Readers in a dock, ship, or a rail yard can use the transmissions from the tags to approximately determine the position of the associated containers. Accurate location information can be obtained if the tag has a built-in GPS receiver.

- **Standards based.** Multiple standards exist for RFID-related applications for freight containers.

Caveats

- **Expensive.** The active tags used for this application member are priced at several dollars per unit. The costs are higher depending on the functionality provided by the tag. For example, a tag might have in-built sensor for explosive chemicals detection and/or the capability to communicate through satellite and function as a GPS receiver to transmit its position in real time. Although such cost can be justified when used in defense and military contexts, the cost of such tags might not be justified when used for commercial freight.

Implementation Notes

Generally, active tags in the 433 MHz frequency range are used with a read range between 100 and 300 feet. This band is available for unlicensed operation in several countries. Various sensors can be embedded inside such a tag to provide specific functionality (for example, detecting explosives or radioactive emission).

In April 2004, the *Federal Communications Commission* (FCC) announced two changes to affirm its support for the security of commercial shipping containers and homeland security:

1. The maximum permissible signal level was increased for RFID systems working in the 433.5 to 434.5 MHz frequency range.

2. The transmission duration limit was increased from 1 to 60 seconds.

The following three ISO standards are pertinent to this application member. All of these standards fall under ISO *Technical Committee* (TC) 104, Freight Containers, Subcommittee 4, Working Group 2:

- **ISO 10374.** This is an existing standard for automatic identification of freight containers using RFID. This standard deals with read-only tags operating in the 850 to 950 MHz and 2,400 to 2,500 MHz frequency ranges. Because these frequencies do not work very well for this application type, however, this standard has not been widely accepted or implemented.

- **ISO 18185.** This draft international standard is directed at electronic container seals. This standard involves both passive and active RFID tags.

- **ISO 23359.** This standard includes read/write RFID tags for freight containers.

Installed Base

Several ports around the world have installed RFID technology to provide port security and security of shipped containers. Some example ports in the United States are Los Angeles/Long Beach, New York/New Jersey, Seattle/Tacoma, and Houston. In the international arena, ports that use RFID-related port security applications (SST) include Antwerp and Rotterdam in the Netherlands, and Felixstowe in the United Kingdom. In 2002, three of the world's largest seaport operators—Hutchinson Whampoa, PSA Corp, and P&O Ports—agreed to use RFID in an effort to enhance seaport security.

4.2 Emerging Application Types

The following are some of the application areas of RFID that hold rich potential for the future:

- Anti-counterfeit
- Smart tags

These application types are not listed in any particular order signifying its importance or the degree of its applicability in an RFID context. It is also possible that the benefits and characteristics of two or more applications types can overlap. The following sections discuss these application types in more detail. For each class, at least one concrete example is provided. For each such example, the benefits and caveats are discussed. An implementation note accompanying each such example provides deployment-specific details.

4.2.1 Anti-Counterfeit

A wide range of items is susceptible to counterfeiting. Some of the items most frequently counterfeited are prescription drugs; currency bills; and high-value items such as perfume, electronics, and watches. Billions of dollars of revenue are lost annually as a result of counterfeiting, in addition to the pain and suffering inflicted on unsuspecting users of counterfeit items (drugs, for instance). An accurate estimate of counterfeiting is difficult because it can be hard to detect and investigate. In addition, counterfeiters are getting sophisticated and technologically savvy, making it impossible for a single technology or method to provide the silver bullet for its prevention. RFID can provide a solution to this problem; to be effective, however, several methods must be used in combination. Some of these measures may be nontechnical, such as licensing policies, education and awareness, law enforcement to tackle counterfeiting, and so on.

Member Applications

This application class type has a large overlap with item tracking and tracing. Indeed, an item to be tracked can be viewed as an asset that can be monitored. Some of the important examples of this application type are as follows:

- **Drug anti-counterfeit.** This is discussed in the next subsection.

- **Currency anti-counterfeit.** This is discussed next.

- **Brand-name and luxury item anti-counterfeit.** RFID can be used to prevent counterfeiting of brand name items (sportswear, for example). Unique item codes can be embedded inside authentic sportswear and tracked. When such a sportswear garment surfaces outside an authorized selling zone, the tag can be used to detect who the original reseller was responsible for distributing it to the gray market. In addition, if a sportswear garment does not have a valid RFID tag (for example, damaged or containing an invalid item code), it can be assumed to be a counterfeit. This can also help to automate inventory in a store, replacing the current manual methods. However, tagging an individual item such as sportswear with an RFID tag is fraught with privacy issues.

The next two subsections discuss what are probably the two most important uses of RFID in preventing counterfeiting.

4.2.1.1   Drug Anti-Counterfeit

The *International Federation of Pharmaceutical Manufacturers Association* (IFPMA) estimates that 2 percent of the drugs sold per year globally might be counterfeit. The *World Health Organization* (WHO) estimates this number to be between 5 percent and 8 percent. Based on these estimates and a global drug market of $327 billion annually, the dollar value of counterfeit drugs ranges from $7 billion to $26 billion annually. A drug can be counterfeited in several ways, including the following:

- Missing active ingredients
- Labeled with an extended expiration date
- Labeled with incorrect potency information
- Labeled with a different drug name
- Contaminated with possible lethal impurities
- Diluted

According to a recommendation by the Healthcare Distribution Management Association, a nonprofit organization for drug distributors, RFID (EPC) tags need to be put on cases by 2005 and applied at the unit level by 2007. However, industry-wide adoption might be as long as 10 years away.

One RFID solution to this crisis is to associate a unique item code or an electronic signature (for example, an EPC [see Chapter 10]) with a case or an individual unit (drugs, medical equipment, and so on) so that it can be tracked through the supply chain. Such an *electronic pedigree* could help to detect counterfeit items entering distribution and to handle returns and recalls. Suppose, for example, that for any particular prescription bottle, an EPC tag must be associated with it for authenticity. Then one, and only one, of the following situations is possible:

1. The bottle does not have an EPC tag.

2. The bottle has an EPC tag that does not exist in the company database.

3. The bottle has an EPC tag that exists in the company database and matches the product details (such as the name; expiration date; when the product was manufactured, distributed, and received).

4. The bottle has an EPC tag that exists in the company database but does not match the product details (such as when the product was manufactured, distributed, and received). Suppose, for example, that the expiration date of a drug reads "February 20, 2006," but the data pulled out from the company database indicates an expiration of "March 15, 2005;" in this scenario, either the drug is a counterfeit or its expiration date has been tampered with. In any case, this drug should be discarded.

Clearly, except for situation 3, the bottle is a counterfeit. A basic assumption underlies this solution: The company databases are hosted securely by the pharmaceutical companies in such a way that an unauthenticated entity cannot tamper with them and that only authenticated entities (for example, a human operator or a computer program) with proper permissions can update them.

Benefits

- Besides anti-counterfeiting, this application offers better supply-chain management and inventory control.

- Drug returns can be facilitated by validating the authenticity of the returned drugs and the purchaser. Similarly, drug recalls can be targeted to customers who bought this drug. (It is assumed that the drug's EPC data will be associated with customer data at the time of purchase.)

- A global EPC infrastructure can provide an effective deterrent to the current counterfeiting problem.

Caveats

- For this type of solution to be effective, a comprehensive EPCglobal type of infrastructure is necessary (see Chapter 10). For this to happen on a global scale, massive investment is needed to set up the hardware, software, and network infrastructure (which might take several years to happen).

- Challenges related to tagging every drug type in its various forms of packaging need to be solved. Most probably, specialized EPC tags will be needed to tag several types of drugs. These might also slow down the application adoption time and increase the cost of implementation.

- The strength of EPC (that is, the company database) is also its weakness. If internal personnel or anyone external can tamper with the company database, EPC might *actually aid* in counterfeiting. In this case, counterfeited drugs gain "legal" status just by having the associated spurious data residing in the company database. In addition, simply attaching a valid EPC tag to a drug is not a watertight anti-counterfeit solution (see the following Implementation Notes); the introduction of counterfeit drugs into the distribution system is still possible.

- If a valid EPC tag associated with a drug gets damaged through mishandling, its authenticity cannot be verified. In this case, the drug must be discarded, resulting in lost revenue. Note, however, that RFID tags are generally sturdy and can withstand environmental conditions to a fair extent.

- Government permission might be needed in some cases to tag certain types of drugs or medical equipment. FDA recommendations on RFID might also contradict or make redundant some directives on unit-level bar coding for patient-safety initiatives. These might impact the adoption time.

- A wide range of personnel must be trained to use the application.

Implementation Notes

Passive tags in the HF (13.56 MHz), UHF (868–870, 902–928 MHz), and microwave (2.45 GHz) frequencies have the potential to be used for this application type. Pallets and cases most probably will be tagged with UHF frequency tags. The actual drugs can be tagged with HF, UHF, and microwave tags. It might seem like a counterfeit drug can be made "valid" by stripping off a valid tag from a valid drug and putting it on this counterfeited drug. However, the tags will be manufactured in such a way that they will break if such an attempt is made, making it evident that some kind of tampering occurred. How about reading the valid unique electronic signature (such as the EPC number) from a legitimate drug, creating a tag with this data, and putting it on a counterfeit drug?

Consider a concrete example. Suppose that a drug called *foobarlene* is being distributed in cylindrical plastic bottles that are 3 inches high, 1 inch in diameter, and weigh 3 ounces. The drug is a pink liquid, and has an expiration date of December 28, 2006. Assume that the valid EPC tag from one of the foobarlene bottles is read by a counterfeiter, and that this person creates and attaches an identical tag using this data to a counterfeited drug. First, the physical characteristics of the drug have to match exactly. For example, the counterfeited drug has to look like a pink liquid packed in a cylindrical bottle 3 inches high and 1 inch in diameter, and must weigh 3 ounces. The bottle must have an expiration date of December 28, 2006. That way, even if someone compares the physical characteristics of this drug, using the EPC tag data, as stored in the company database, those characteristics will match with the counterfeit one. So comparing the physical characteristics in this case is not sufficient to determine conclusively whether this drug is indeed a counterfeit.

Next, the intended distribution region, as stored in the company database for this drug bottle, is compared with the counterfeit one. Suppose, for this example, that the drug was intended for distribution in South Africa, but actually shows up in China instead. Clearly, something is wrong, and this bottle of foobarlene is a counterfeit suspect. However, what happens if this drug shows up in South Africa instead? Because the intended destinations match, it cannot be determined for sure that this is a counterfeit drug (although in this case, it is). Therefore, if on its way to South Africa, this bottle of foobarlene is replaced with a counterfeit one that has exactly the same physical characteristics of that of the actual one, the company product data associated with the EPC tag will *not* be sufficient to determine conclusively that this drug is a counterfeit. Some other measures will be necessary (for example, opening a bottle of foobarlene randomly and testing the chemical properties of the drug to see whether it matches the composition specified in the company database).

As you can understand from the preceding example, one of the soft spots of this unique electronic signature solution seems to be between tag reads. That is, the counterfeits can be introduced by replacing the valid drugs between tag reads (for example, between two valid distribution points). Counterfeit drugs can be introduced, for instance, during transit between two valid distribution points, by replacing a valid lot of drugs with counterfeit drugs having the same EPC tag data, the same EPC case and pallet tag data, and the same physical characteristics as the replaced lot. Admittedly, pulling this feat off will require some very sophisticated and resourceful counterfeiters. In this case, the solution can do very little to identify these counterfeit drugs. The stolen valid drugs can then be sold on the black market to individuals who are not at all concerned whether a bottle of this stolen drug has a unique electronic signature. Note that in this case, no tampering with the company database is needed.

Dishonest sellers can also abuse electronic signatures to purvey counterfeit items. In one case, the seller can just put duplicate electronic signatures (for example, EPCs) on counterfeit items and bypass recording the sale of any of these counterfeits (as well as bypass the recording of the sale of the original item that had the genuine EPC associated with it). As a result, a duped customer can "authenticate" the counterfeit item's EPC from reliable sources (except the item information might indicate that it has not sold yet!) and might think that he indeed bought a genuine item. In another simple scheme, the seller of a counterfeit item can scan an EPC of a counterfeit item and show the customer the associated electronic pedigree and product-specific data on his computer to verify its "authenticity." However, instead of getting this data from the manufacturer/product database, the seller can spoof it locally using a simple program on his computer!

The customer might not be able to recognize this ploy easily because the header, manager number, and object class of the counterfeited item's EPC, together with valid product information and the electronic pedigree, can easily be copied from a valid EPC of a similar product. In this case, the buyer might be easily convinced that he is indeed buying a genuine product! The seller can create further confusion, if confronted (after the buyer confirms it is a counterfeit via a valid source), by accusing technology glitches. ("Didn't I show you that the EPC was valid? How can it be invalid now? Let me show it to you again. That's the problem with this technology—it doesn't work all the time, and look who is getting blamed for using it!")

These scenarios clearly do *not* represent a weakness of the technology or the EPC scheme, because parts of the technology are selectively ignored to misuse and misrepresent the technology to sell counterfeit items. These scenarios do show that security and anti-counterfeiting are not simple issues that can be solved just by attaching RFID tags to items. This is where nontechnical measures, such as tracking down and prosecuting counterfeit sellers and black marketeers, strict control of transportation, physical security of the transported lot, and so on, are necessary to ensure that the technology delivers its potential benefits to the customers and end users. Then, the RFID solution coupled with these nontechnical measures will provide a solution that is close to bulletproof.

4.2.1.2 Currency Anti-Counterfeit

Currency counterfeiting is one of the oldest crimes in history. According to one U.S. Secret Service official, about $63 million in U.S. counterfeit currency was seized in 2003, of which $10.7 million was seized in the United States.[6] Columbia accounted for more than $31 million, being the single largest producer of U.S. counterfeit currency. The Secret Service estimates that more than 42 percent of counterfeit currency passed domestically in 2003 was produced outside of the United States, whereas 46 percent of this amount was produced in the United States. Organized crime networks exploit the opportunities associated with "dollarization," which refers to the process by which a foreign country (for example, Ecuador) adopts the U.S. dollar as its national currency. These countries are prime targets for counterfeit U.S. currency. In the United States, an individual is *solely* responsible for the authenticity of the currency he carries. Authorities can seize counterfeit currency from an individual without any legal requirement to compensate this person. Likewise, the financial institution that dispensed the counterfeit currency to this individual is also *not* legally liable to compensate him.

RFID can be used to provide authenticity of paper currency bills. Very small RFID tags can be concealed in a currency bill. Such a tag can carry EPC data or some kind of unique identifier that can be read by special readers. If a tag is absent or if its data cannot be matched against the currency database, it can be assumed to be a counterfeit. Multiple tags, each carrying its unique data, can be inserted into a single currency bill so that the combination of the data from these tags

---

[6] Statement of Mr. Bruce A. Townsend, deputy assistant director, United States Secret Service, before the Committee on Financial Services, April 28, 2004.

uniquely identifies and authenticates this currency bill. This, in turn, will also make the counterfeiting difficult.

Benefits

- **Thwart nonexpert counterfeiters.** An RFID solution would eradicate the vast majority of current and would-be counterfeiters who are not technically savvy.
- **Track illegal transactions.** A currency bill can be tracked at a transaction level, which makes this solution an effective deterrent of illegal transactions. For example, when an ATM dispenses cash, the tag data of the currency can be read and associated with the account from which it is withdrawn. Similarly, when a currency note is deposited in the bank, the tag data can be associated with the account into which it is deposited. In both of these cases, the person(s) holding the account can be associated with handling this currency note.
- **Effective way to track blackmailers.** The currency notes can be read and marked in the database before these are handed out to blackmailers whose identities cannot be disclosed at the transaction time. Next time the current notes are traced (from the previous benefit), the data can be used to identify the blackmailers.

Caveats

- For this type of solution to be effective, a comprehensive EPCglobal type of infrastructure is necessary (see Chapter 10). For this to happen on a global scale, massive investment is needed to set up the hardware, software, and network infrastructure (which might take several years to happen).
- If a valid EPC tag associated with a currency note is damaged through mishandling, its authenticity cannot be verified. In this case, the note might have to be marked as destroyed. Note, however, that RFID tags are generally sturdy and can withstand environmental conditions to a fair extent.
- Associating currency note EPC tag data with a consumer records might raise privacy concerns.
- Undue harassment of innocent persons might result as a side effect of associating transactions with a currency note. Suppose, for example, that a person A receives a valid currency note from a bank ATM (which records the tag ID of this currency note with this person's account). Assume that this note is subsequently counterfeited, without any knowledge of person A, and the authorities track the counterfeited note. Because the last transaction using the valid currency is associated with person A, he might become a target of the counterfeiting investigation. In another situation, person A uses this currency note to pay his taxi fare to the airport. The taxi driver is subsequently robbed, and the

robber uses the currency bill to pay for drugs. The police catch the drug seller and retrieve the currency note. Person A could be in deep trouble now! The main issue in the second example is that not all cash transactions can be recorded to link the currency note with a transaction. These gaps can reduce the effectiveness of such tracking applications to a major extent.

- A wide range of personnel must be trained to use the application properly.
- Finally, the tag cost has to come down substantially from where it is today.

Implementation Notes

The possibility of tampering with the currency database, although theoretically possible, is remote. After all, it would probably be among the most secure databases in existence. Passive tags in the microwave (2.45 GHz) frequency range will be used because these require only small antennas to communicate with the reader (a crucial element that allows a tag to have a very small form factor so that it can be embedded in paper currency).

4.2.2   Smart Tags

The *smart tags* application type almost seems like an unfair catchall classification of an emerging application type! A smart tag is essentially an active (or a semi-passive) tag that has a battery and on-board electronics, and can therefore perform custom tasks besides just storing and transmitting data unique to an attached object. The key phrase here is *custom tasks*, which can be *anything*! For example, a custom tag can monitor and report its surrounding temperature, humidity, radioactive emissions, and so forth, among other unlimited task types. In other words, RFID can be combined with sensor technology to create a variety of smart tags. The application members belonging to this application type are only limited by one's imagination. Note that a smart tag can be of any physical dimensions as long as it can be deployed. It can be the size of a pack of cards, laptop, or a suitcase, provided it can be attached properly to the targeted object. Therefore, "anything" seems to be fair game when it comes to smart tags! It is the author's personal opinion that this application type holds the most potential for RFID technology of the future.

Member Applications

These application types do not exist today. Several of the application members (current and future) that belong to this application type may also belong to a different application type. For example, today, the electronic toll collection system uses smart (semi-active) RFID tags that can indicate whether the account balance of a person's toll account is at an acceptable level. This application is thus a member of both smart tag and electronic payment application types. Some example applications belonging to this type are as follows:

- **Toll payment.** This is described earlier in the discussion of prevalent application types.
- **Smart expiration-date determination.** Every perishable item has an expiration date that is determined at the time of its manufacture by assuming certain storage conditions,

such as temperature, humidity, and so on. In reality, however, the "safe" consumption period can vary considerably. Consider meal packets to be consumed by soldiers in the field. Assume that these packets are stored in local storage areas near battlegrounds. The environmental conditions of these storage areas might be unpredictable. For example, a storage area might be located in the middle of a desert or in a high-altitude mountain cave, where temperature and humidity variations are extreme. As a result, the original expiration dates that were calculated based on average environmental conditions are not valid anymore. In this situation, smart tags attached to cases of such packets can determine environmental changes and can automatically calculate new expiration dates. These tags can then transmit this information to a central inventory system, which can then direct army personnel to discard meal packets whose new expiration dates have passed. The whole process of determination and notification is automatic and accurate.

- **Smart weapon.** This is discussed in the next section.

Because of the high number of members of this application type and their substantial variations, the following section covers a sample member (a smart weapon) to provide a glimpse of the potential of this application type.

4.2.2.1 Smart Weapons

Suppose that an army is at war with enemy forces in hostile terrain.[7] It is difficult to send a large number of forces through unexplored and dangerous terrain to secure it. It is equally difficult to keep the terrain impassable for enemy forces. However, control of the terrain might prove very crucial to winning the war. Currently, land mines are used to make a terrain inhospitable to enemy forces. However, land mines make the terrain equally dangerous for both forces. In today's high-technology warfare, military planners want a "smarter" solution.

Enter RFID active tags used to design a new breed of smart weapons that can recognize a friendly force from an enemy force and can modify their behaviors accordingly. The military can airdrop a large number (perhaps several thousand) of these smart weapons (for example, high explosives), with an active RFID tag attached to each, on a particular terrain to make it inaccessible. A friendly force passing through this terrain can instruct, securely, the RFID active tags attached to these weapons to deactivate the weapons. After passing through the terrain, the force can, conversely, turn the weapons back on, rendering the terrain inhospitable to the enemy force.

The possibilities do not stop here. These weapons could, in real time, sense the external conditions of the terrain (for example, vibration of the ground or a loud sound) to signal the approaching enemy, and relay the information securely back to the command center. The command center can then instruct these weapons to turn on their more sophisticated features to fine-tune the monitoring and action (for example, specify conditions when these weapons can detonate). In addition, such a tag can communicate with other similar tags, forming a huge ad hoc

---

[7] The author does not support or condone war in any form. This is for illustration purposes only.

wireless network (for example, a sensor network) to exchange data among themselves. For example, a shutdown command received by some of the smart bombs at the edge of the network can be transmitted to the remaining nodes, even if they are located deep inside. In other words, these smart weapons can replace human soldiers when the situation is too risky or the requirements are so demanding that it is impossible or very difficult for a human to perform the tasks satisfactorily for a long period of time in a continuous manner.

Benefits

- **Flexible.** Smart weapon feature combinations are virtually unlimited.
- **Effective.** A smart weapon, unlike a human, does not tire (unless, of course, the battery runs out) when performing tasks continuously. Even if enemy forces capture such a weapon, they cannot use it as a negotiating chip.
- **Possible to mass produce quickly.** The production of smart weapons is only limited by the production capacity of the manufacturers.

Caveats

- **Could be expensive.** Depending on the type of features supported by the active tag of a smart weapon, the cost of producing these tags will vary. More features generally means more complexity, and hence a higher production cost.
- **Extensive field testing required.** The active tags for smart weapons must be field tested thoroughly under different operating conditions to make sure they function reliably, which increases implementation time and cost.

Implementation Notes

Smart tags will most likely operate at a 2.45 GHz or 5.8 GHz frequency range and be equipped with a battery (typically, a 5-year battery) and have specialized on-board electronics geared toward performing specialized tasks. A typical read distance of such a tag will be more than 100 feet (about 30.5 meters). Typical data transmission rate (to its surroundings, irrespective of the presence of a reader) will be once every few seconds to once every several hours. On the other hand, a smart tag could also be a semi-active tag that can be made to transmit its data in the presence of a suitable reader. A smart tag will most probably have a variety of built-in sensors to monitor its surroundings and be able to communicate with other similar tags (thus forming an ad hoc wireless network).

4.3 Conclusion

Fueled by the rapid rate of advancement of RFID technology and the products available, the range of RFID application types is continuously expanding. Some application types are already mature and being used commercially; other promising types are currently in the prototype stage. Several of these types might be commercially deployed in the future, depending on the results of the prototypes, the willingness of the business community to invest capital and undertake the risks, and consumer and user acceptance. This chapter covered some of these important application types from several aspects. The types discussed here are by no means exhaustive. By the time you read this book, other application types might exist that were not available at the time of writing; this is especially true regarding emerging application types. However, from the information you have learned in this chapter regarding the important types and their members, you should have no problem understanding and classifying and unfamiliar ones as you are exposed to them.

Privacy Concerns

Not surprisingly, privacy issues represent a significant concern regarding RFID technology. After all, when a new technology is invented and is in the process of being developed, it must be analyzed from several viewpoints to determine whether and how its use will impact society. Consumer advocacy groups worry that RFID misuse might lead to the tracking of individuals, resulting in a loss of privacy.

In the midst of all the heated discussions about RFID and privacy, one important point is that privacy issues apply only to some particular types of RFID applications. A large number of RFID applications raise no privacy concerns. With regard to those RFID applications that do raise security concerns, the RFID vendors, the business community, researchers, integrators, lawmakers, and privacy advocates are feverishly working to deal with these issues. However, agreed-upon standards and regulations will eventually result from this meeting of the minds.

CAVEAT EMPTOR

Neither this chapter nor this book provides legal counsel. Adherence to the advice provided herein does not guarantee compliance with any law or standard, domestic or international. Example rules and regulations in this chapter and book are just that: examples. Businesses should consult their own legal counsel to ensure compliance with laws and regulations that might apply to them.

5.1   Core Issue

By its very nature, RFID technology can identify almost any type of object, even down to an individual component level. For example, *any* T-shirt produced by *any* manufacturer in the world can be *uniquely* identified using RFID. An identifier scheme such as *Electronic Product Code* (EPC;

see Chapter 10, "Standards") makes it possible to generate a large number of unique identification numbers. Each of these unique identification numbers can be put on an RFID tag, which can then be attached to each item of a particular type. Continuing with the T-shirt example, it is possible, using an EPC numbering scheme of appropriate size, to tag every T-shirt produced in the world in any given year. The tag might be hidden or embedded in a T-shirt in such a manner a customer cannot find it. When a customer buys such a tagged T-shirt, this unique ID can be recorded at the time of sale and associated with the customer's personal record. When this customer carries or wears this T-shirt, a concealed reader can, theoretically, read the tag "anywhere," "anytime," without this person's knowledge or consent. The tag data can then be used by some kind of application to extract the associated personal record, resulting in tracking of this item and hence its owner. Of course, this scenario assumes that the tag is not destroyed before the customer leaves the store, his personal profile is somehow accessible and associated with the tag's EPC identifier at the time of purchase, and that some kind of massive distributed database exists that can store and update the data of each such T-shirt EPC and each customer's personal information. The purchaser-wearer of this T-shirt loses anonymity and control of how the collecting parties use this tracking information, which can result in uncontrolled profiling of this customer and might seriously infringe upon his privacy rights.

Privacy-rights advocates fear that if the use of RFID is not checked, its use might open doors to the government, law enforcement officials, business community, and criminals alike to surreptitiously read the unique tag data of items a person wears or carries (in a purse, for example, assuming the purse is made of an RF-friendly material and is in the read range, among other things; see Chapter 2, "Advantages of the Technology") and extract the information of the items and buyer associated with this data. Thus, the buyer loses his privacy and anonymity. Moving closer to what is reality today, you need to understand that RFID has severe limitations in terms of reading distances (depending on the frequency and tag type) and materials through which it can read tag data, to name a few (see Chapter 3, "Limitations of the Technology"). Therefore, currently, RFID cannot read a tag at an arbitrary distance through an arbitrary material in an arbitrary operating environment.

In this discussion, one subtle but important point is that two *distinct* elements are involved in item-level tagging: unique tag data and *consumer identification data* (CID) of the customer. As long as these two entities are kept separate, the question of privacy infringement of the consumer does not arise. When these two pieces of data are linked or associated somehow, however, *then* the issue of privacy-rights violation might arise. To alleviate this problem and build customer trust and confidence, a retailer might want to explain to customers the benefits of item-level tagging. It can then offer an opt-in or opt-out option to customers for collection and association of their CIDs with purchased items' tag data.

Note that privacy concerns already exist with bar code, credit card, and consumer discount card technologies, and that RFID is another variation on these same themes. Retailers can already tie together data from credit cards and customer bar code loyalty cards, which raises some of the same privacy issues as RFID.

However, not all item-level tagging involves privacy concerns. Applications that involve electronic payment are acceptable, for example, as is the tagging of a shipping container with the sender's information (see Chapter 4, "Application Areas").

5.2 What Are Privacy Rights?

Privacy rights vary from country to country. "The Right to Privacy," a Harvard Law Review article by Samuel D. Warren and Louis D. Brandeis, is probably the most fundamental work regarding individual privacy rights in the United States.[1] Early in this article, the authors describe the concept of privacy as follows:

> That the individual shall have full protection in person and in property is a principle as old as the common law; but it has been found necessary from time to time to define anew the exact nature and extent of such protection. Political, social, and economic changes entail the recognition of new rights, and the common law, in its eternal youth, grows to meet the new demands of society. Thus, in very early times, the law gave a remedy only for physical interference with life and property, for trespasses vi et armis. Then the "right to life" served only to protect the subject from battery in its various forms; liberty meant freedom from actual restraint; and the right to property secured to the individual his lands and his cattle. Later, there came a recognition of man's spiritual nature, of his feelings and his intellect. Gradually the scope of these legal rights broadened; and now the right to life has come to mean the right to enjoy life, —the right to be let alone; the right to liberty secures the exercise of extensive civil privileges; and the term "property" has grown to comprise every form of possession—intangible, as well as tangible.

The article then analyzes the issues involved with privacy, and states the following:

> The protection afforded to thoughts, sentiments, and emotions, expressed through the medium of writing or of the arts, so far as it consists in preventing publication, is merely an instance of the enforcement of the more general right of the individual to be let alone. It is like the right not be assaulted or beaten, the right not be imprisoned, the right not to be maliciously prosecuted, the right not to be defamed.

In essence, the right to privacy is the right "to be let alone," and is considered to be among the basic rights of an individual. One way to look at privacy rights of an individual is as a personal copyright that others cannot publish indiscriminately. The law review article also analyzes the limitations of the right to privacy, of which the two most significant ones are as follows:

- Publication of matter that is of general and public interest
- Publication of the facts either by the individual or with his consent

---

[1] "The Right to Privacy." Samuel D. Warren and Louis D. Brandeis. *Harvard Law Review*, Volume IV, December 15, 1890, No. 5.

If an individual publishes the facts himself or permits a third party to publish his personal facts, the publication of such does not constitute a violation of the privacy rights of this individual. Any publication (which may involve sharing of information with other parties, either oral or written, irrespective of whether it is recorded) of the personal facts of a person without his consent almost always constitutes a violation of his privacy rights.

5.3   Resolution Attempts

Current attempts to assuage privacy concerns posed by RFID generally fall into the following three categories:

- **Political and legal (governmental).** Politicians, lobbyists, and privacy-rights advocacy groups are attempting to enact legislation (and guidelines) regarding RFID use so that it does not pose a threat to the privacy rights of individuals.
- **Business community.** This second category represents the second most crucial element in the privacy equation after consumers—the business community. Businesses can proactively take effective measures against privacy-infringement issues as an independent effort.
- **Technology community.** Developers of RFID technology and related products are attempting to provide solutions to prevent the unauthorized use of RFID tags to snoop on individuals.

Within each of these categories, and sometimes overlapping, people are working on solutions to provide a satisfactory resolution to the privacy issues RFID raises. The following section discusses some of their current activities.

5.3.1   Political and Legal

Senator Patrick Leahy (D-VT) has expressed concerns that federal laws might be necessary before the privacy-infringement issues of RFID go too far.[2] Senator Leahy proposed that answers should be sought on what information is collected; how it is collected, stored, accessed, secured, corrected in case of mistakes; and conditions under which it can be used by law-enforcement agencies. He also acknowledged that it is important to let RFID mature without needless roadblocks.

In November 2003, eight privacy-rights advocates, including *Consumers Against Supermarket Privacy Invasion and Numbering* (CASPIAN), the *American Civil Liberties Union* (ACLU), and the *Privacy Rights Clearinghouse* (PRC), issued a privacy position statement. This statement proposed the following three-part framework for the proper use of RFID:

---

[2]   Remarks of Senator Patrick Leahy. "The Dawn of Micro Monitoring: Its Promise and Its Challenges to Privacy and Security." Georgetown University Law Center, March 23, 2004.

- A formal technology assessment
- Adherence to a set of proposed principles of fair information practices
- Outright rejection of certain types of uses of RFID

In addition, it is also requested that manufacturers and retailers impose a voluntary moratorium on unit-level RFID tagging of consumer items until a formal assessment of the technology by all stakeholders and consumers is performed.

In Missouri, SB 0867, otherwise known as the RFID Right to Know Act of 2004, was introduced by State Senator Maida Coleman (D) on January 7, 2004. This bill requires that an RFID tag on an item have a clearly visible label that explicitly mentions that the item contains an RFID tag and that the tag can transmit unique identification data to a reader both before and after the sale of the item. The bill failed to pass when its hearing was cancelled on March 9, 2004.

On January 28, 2004, the Radio Frequency Identification Right to Know Act (HB 251) was introduced in Utah and sponsored by David L. Hogue (R). It was initially approved by the Utah house of representatives and the Utah senate's Business and Labor Committee. One amendment to this bill would have required retailers to destroy a tag unless they notified consumers of its existence and capabilities. This amendment seemed unpopular with retail associations. However, the bill expired in March 2004, before the Utah senate could vote on it. It is expected that this bill will be reintroduced.

In April 2004, the California state senate approved bill SB 1834, introduced by Senator Debra Bowen (D), to impose regulations on the uses of RFID by libraries, retailers, and other private bodies. The bill proposed the following three rules for acceptable use of RFID by a business to collect data related to personal identity:

- Inform customers whenever RFID is used in items to collect data.
- Customer permission is mandatory before his purchases can be tracked.
- All RFID tags must be deactivated before a customer leaves the store. (This was subsequently modified.)

This bill was defeated by the members of the California State Assembly on June 25, 2004.

In June 2004, an RFID Privacy Guideline statement was jointly released by the *Ministry of Economy, Trade, and Industry* (METI) and the Japanese Ministry of Public Management, Home Affairs, Posts and Telecommunications. Some of the salient points of these guidelines are as follows:

- Notify that RFID tags are attached to items.
- Provide consumers with an opt-out policy with regard to linking the RFID tags to their personal information. The guide proposes informing customers how to defeat RFID technology so that the tags cannot be read (for example, using metal foil to cover the tag or physically removing it from the item).
- Restrict the collection and use of information when private data is stored on tags.

- Ensure information accuracy when tags carry private information.
- Share information with the customers.

Privacy-advocacy groups strongly urged discussion at the federal level regarding large-scale use of RFID by retailers and the government. The U.S. *Federal Trade Commission* (FTC) responded by hosting an RFID workshop in June 2004. At this workshop, the privacy-advocacy groups called for the FTC and other governmental agencies to conduct an impartial assessment of RFID technology. The supporters of RFID expressed concerns that such an assessment might be too premature. FTC has prepared a report based on its research and findings at this June workshop.[3] This report calls for self-regulation of the manufacturers and users of the RFID technology. At the same time, FTC has not completely ruled out issuing future RFID guidelines.

On January 19, 2005, Article 29 Working Party, the European Union's advisory on privacy and data protection, published a report that provided privacy guidelines regarding RFID use. The report called for obtaining clear consent from an individual when RFID is used and making him aware of the following:

- The presence of RFID tags in the merchandise
- What type of personal data is collected and how it is processed
- The right to access and check accuracy of personal information collected by a business

In March 2005, the New Mexico house of representatives Judiciary Committee dismissed the bill HB215, titled Removal of Radio Frequency ID Tags, sponsored by representative Mimi Stewart (D). This bill called for attaching a label with a tagged item clearly stating that the item contains an RFID tag. The bill proposed that all tags be removed from items before they leave the store and that a business should provide a customer, upon a written request, all personal information that it has collected.

In May 2005, Massachusetts State Senator Jarrett Barrios (D) introduced bill No. 181, a bill similar to SB 1834, in the state senate. However, Senator Barrios did not support the deactivation or "kill" feature for a tag. His concern was that if a tag is deactivated, useful information required for recycling (for example) might be unavailable for such a tagged item. He also expressed a concern that laws that enacted today might not be applicable in the future because of the evolving nature of RFID.

5.3.1.1   Privacy-Rights Advocacy Groups

Several privacy-rights advocacy groups are involved in the RFID privacy debate with the supporters of the technology. Some of the most prominent among these advocacy groups are the following:

---

[3]   Radio Frequency Identification: Applications and Implications for Consumers. A Workshop Report from the Staff of the Federal Trade Commission, March 2005.

- **CASPIAN** (Consumers Against Supermarket Privacy Invasion and Numbering; http://www.nocards.org/, http://www.spychips.com/) organizes and participates in privacy-related events and debates and publishes privacy-related policy documents.

- The **ACLU** (American Civil Liberties Union; http://www.aclu.org/) champions freedom-of-speech rights, equal protection rights, due process rights, and last, but certainly not least, privacy rights.

- The **PRC** (Privacy Rights Clearinghouse; http://www.privacyrights.org/) publishes information about protecting consumer privacy rights.

- The **IAPP** (International Association of Privacy Professionals; http://www.privacyassociation.org/) is an association of privacy and security professionals that provides interaction, education, and discussion regarding privacy matters.

- The **EFF** (Electronic Frontier Foundation; http://www.eff.org/) is an organization devoted to protecting the fundamental rights of individuals and raising and debating the civil liberties issues associated with technology.

- **EPIC** (Electronic Privacy Information Center; http://www.epic.org/) is a public interest research center whose primary goals are to protect privacy and bring public attention to emerging civil liberties issues.

5.3.2 Business Community

Instead of solely depending on the government and technology community to resolve RFID privacy concerns, the business community has an equally crucial role to play. Direction and guidelines have already started to emerge. For example, EPCglobal provides a set of guidelines targeted at the business community that relate to the use of EPC in consumer products.[4] The following list provides sample general guidelines (strictly for informational purposes only):

- Unambiguously document the corporate privacy policy.

- Publicly share this policy with employees and customers.

- Implement an effective feedback system that captures and resolves concerns related to the published corporate policy.

- Clearly state on tagged item packaging the presence of RFID tags in the merchandise, and make customers aware of RFID use in the store.

- Inform the customers as to what data is stored on the tags and how it is going to be used and the benefits (both for the business and the customer) associated with it.

- Provide customers the option to either accept or reject associating their CID with the purchased item tag data. If the customer opts out, disable or remove the tag at the point of sale.

---

[4] Guidelines on EPC for consumer products (http://www.epcglobalinc.org/public_policy/public_policy_guidelines.html).

- Educate customers about the capabilities and limitations of RFID.

- Implement proper record keeping and security measures to retain and secure generated data (including end-to-end security from the tag readers, underlying networks to the databases that store the final data, and so on).

Communication, compliance, and consistency regarding RFID use are represented in the preceding guidelines; such guidelines will help businesses preserve customer goodwill and trust. However, guidelines such as these might impact existing business processes and, therefore, call for careful planning and execution.

5.3.3   Technical Community

While political, legal, and business interests engage intensely regarding RFID use, the technology researchers and vendors are also working feverishly to provide technical solutions to RFID privacy concerns. Essentially, these solutions are broadly targeted at deactivating (or "killing") a tag or paralyzing the capability of readers to read a tag (assuming it is "alive").

The subsections discuss these solutions in detail.

5.3.3.1   Kill Commands

This kill mechanism is targeted at the tag itself to render it useless. Originally developed by MIT's Auto-ID Center research group in 2003 for its EPC specification (see Chapter 10), this idea seems to have the support of several privacy-rights advocacy groups and state legislators. The idea calls for a reader command, called a *kill command*, which, when issued to a live tag, instructs the tag to self-destruct. When a tag receives such a command, the tag can erase its memory or reconfigure itself in such a manner that it cannot communicate any further with a reader. A password can also be associated with a kill command so that a reader can securely issue this command.

Early prototypes of chips implementing this command were developed in 2003. Alien Technology, Inc., developed an Auto-ID Class 1 UHF specification chip that implemented a prototype version of this command. Similarly, Matrics developed a prototype chip based on the Auto-ID Class 0 UHF specification, and Philips Semiconductors built a prototype chip based on a 13.56 MHz standard. The implementation of this command seems to be simple and is not expected to increase the unit price of a tag. The chief drawback of this approach is that the original tag is destroyed; therefore, the potential to use the associated data dies with it. This total "kill" might not be desirable in some cases. Suppose, for example, that a consumer buys a toxic substance that has an RFID tag that contains recycling information for this product. If the consumer requests that this tag be destroyed at the time of purchase, the recycling information specific to this product cannot be accessed anymore (which might lead to unwanted consequences such as the dumping of this product in a landfill and so on).

5.3.3.2 Blocker Tags

A blocker tag is a simple, yet ingenious, mechanism targeted at the RFID reader to render it use-less in communicating with tags in its read zone.[5] However, the tags are not destroyed and may remain alive. Originally developed by Ari Juels, Ronald Rivest, and Michael Szydlo, this idea calls for a unique tag called a *blocker tag* that masquerades as a valid tag with some special prop-erties. The readers, which use what is known as the *tree-walking singulation algorithm* to read unique tag data, are the targets of this blocker tag. This algorithm is used by virtually all the read-ers in the UHF frequency and, therefore, is the most effective in blocking these types of readers. The blocker tag mechanism can also be implemented for the ALOHA algorithm, which is chiefly used by all 13.56 MHz frequency readers. For you to understand how a blocker tag works, you must understand the workings of the tree-walking singulation algorithm. The following example describes this.

Suppose that three persons, each with a three-letter name (in this case, Bax, Bob, and Tim) are standing in front of a blindfolded interrogator. The interrogator knows neither how many per-sons are standing before him nor what their names are; he does know that each of their names consist of three letters only, however. The interrogator is tasked with discovering the names of all those standing before him. He can only deduce the number of persons present from the number of unique names he can determine using the letters from English alphabet. He can only accept a response from one person at a time; if more than one person tries to talk to him, a *collision* results. The interrogator must resolve a collision whenever it occurs. The interrogator can ask these per-sons to selectively tell him parts of their names if those parts match his questioning criterion. A person must remain silent if his name does not satisfy this criterion. Using this scheme, the inter-rogator can discover the names of these three persons.

The interrogator first asks those whose name starts with an *A* to respond; the rest should remain silent. No one answers. The interrogator asks them to respond if the first letter of their name starts with *B*. Bax and Bob both respond, resulting in a collision. The interrogator then asks those whose name starts with *Ba* to respond. Only Bax answers. The interrogator then asks for those whose name starts with Baa to respond; they all remain silent. The interrogator continues with name prefixes (Bab, Bac, and so on) until he arrives at *Bax*. At that point, only Bax responds. The interrogator has now found out the name of one of the persons. The interrogator then repeats the whole set of commands to the persons standing before him, starting with *Bb*, *Bc*, and so on. He only gets an answer from Bob when he uses the name prefix *Bo*. The interrogator then repeats the same procedure starting with name prefix *Boa*. Again, he gets an answer from Bob only when he arrives at Bob. Now the interrogator has successfully found out the names of two of the per-sons standing before him. He then proceeds to repeat the entire process with the *C*, *D*, and so on, until he calls out *T*. At that point, only Tim answers. The interrogator delves into name prefixes

---

[5] *The Blocker Tag: Selective Blocking of RFID Tags for Consumer Privacy*. Ari Juels, Ronald L. Rivest, and Michael Szydlo. V. Atluri, ed. 8th ACM Conference on Computer and Communications Security, pp. 103–111. ACM Press. 2003.

starting with *Ta*, *Tb*, and so on. He gets a response from Tim when he arrives at *Ti*. The interrogator then starts from the *Tia*, and receives a reply from Tim when he gets to *Tim*. Thus, the interrogator discovers all three names. The interrogator then asks those standing before him to respond if their name starts with *U*. No answer. *V*. No answer. The interrogator continues with no answer through *Z*. At that point, the interrogator knows that three persons are standing before him and that he has discovered each of their names.

Now coming back to blocker tags, suppose that there is now a fourth person, and this person does not have any single unique name but instead has every three-letter name possible (that is, can arrange his name in *any* three-letter combination). In essence, this is what defines a blocker tag. A blocker tag is a kind of *super* tag that can assume any value of a tag allowed in the range of possible values. Suppose that this fourth person can speak in such a manner that it seems to the interrogator that two different persons are responding to him at the same time, resulting in a collision. Let's see how this property can totally confuse the interrogator.

When the interrogator starts with *A*, Bax, Bob, and Tim remain silent; however, the fourth person responds, resulting in a collision (because he speaks in a manner that seems that two persons are speaking). To resolve this situation, the interrogator moves to *Aa*, but again this fourth person speaks out, resulting in another collision. The interrogator moves to *Aaa*, and gets a response from the fourth person. The interrogator assumes a person named *Aaa* is present, and then moves on to *Aab*; the fourth person responds again. The interrogator is forced to explore every three-letter name possible. However, for three-letter names, the possible number of three-letter names is $26 \times 26 \times 26$ (or 17,576; not exactly a small number). For names with more letters, the situation gets worse exponentially; in that scenario, the interrogator becomes overwhelmed after some time and stops.

In the case of a tag, the "letters" are from the binary alphabet and can be either 0 or 1. The interrogator is a reader that uses a command set to selectively elicit response from the tags. Tag "name" lengths typically start from 96 bits in the current tags, which results in a namespace size of about 80,000 trillion trillion! A reader will spin its wheels forever trying to find this many unique tags in its read zones when a blocker is present. Even if a reader can determine all these tags in its read zone, it does not mean all these tags are valid; therefore, a lot of the reader's time and effort are wasted in determining unnecessary data. In reality, a reader becomes overwhelmed after a few thousand tries and stops. Thus, in effect, a blocker tag does what its name suggests: It blocks a reader out completely so that it cannot read *any* tag in its read zone.

Clearly, a blocker tag needs two antennas to transmit two responses (0 and 1) at the same time to a reader. The first working prototype of a blocker tag was demonstrated at the thirteenth annual RSA Conference in San Francisco, California, in February 2004. It is expected that a production version of blocker tags will be available sometime in 2005.

A concern with blocker tags is that they can be used maliciously to cripple the operations of a business (a warehouse operation, for example).

5.4 Conclusion

The privacy implications of RFID are being hotly debated today in political, legal, business, and technical forums. Interested parties in the debate include the RFID technology backers, privacy-rights advocates, businesses, politicians, researchers, and consumers. Because RFID is an emerging technology, however, laws enacted now might become obsolete in the future. On the technical front, both tag-level and reader-level schemes have been invented in the form of kill commands and blocker tags, respectively, to deactivate the readability of a tag.

A combination of legal and technical measures might offer a solution to the privacy issues that can be broadly accepted by all the interested parties. Although a clear consensus has yet to emerge, on the political and legal front, support is gathering around a three-point standard of informing customers of the presence of RFID tags in purchased items, obtaining customer permission to use the tag data, and destroying (optionally) the tags before customers leave the store.

CHAPTER 6

RFID Versus Bar Code

RFID is currently being touted as a "better bar code" and "smart bar code." The media regularly proclaims that the days of bar code are numbered and that RFID will replace bar codes "soon." In fact, RFID does have some clear-cut advantages over bar code, but bar codes also offer some clear-cut advantages over RFID. Unfortunately, in all the enthusiasm of advertising the advantages of RFID over bar codes, the other side is silently, sometimes deliberately, neglected in the press. As a result, a common belief is taking shape that bar codes are a sure loser when compared to RFID, irrespective of the context. This is a completely wrong belief!

This chapter objectively analyzes the relative advantages and disadvantages of RFID and bar codes. After reading this chapter, you will better appreciate the strengths and weaknesses of these two technologies, which will enable you to choose the appropriate technology based on its actual merits and not how it is viewed today. This chapter reviews material from Chapter 2, "Advantages of the Technology," and Chapter 3, "Limitations of the Technology," and closes with a discussion of what is probably the most popular belief today about RFID: Soon, RFID will completely replace bar codes. Before delving into an analysis of bar codes versus RFID, you need to understand bar code technology itself, which the following section covers. This understanding will help clarify some of the issues discussed in the subsequent sections. If you are already familiar with bar codes, you can skip the following section and go directly to Section 6.2.

6.1 Bar Codes

This section provides a brief introduction to bar code technology. The chief aspects of bar codes are discussed together with some of the most popular standards in use today.

6.1.1   What Is a Bar Code?

A *bar code* is a scheme in which printed symbols represent textual information. The printed symbols generally consist of vertical bars, spaces, and squares and dots. A method that encodes alphanumeric characters using these symbol elements to a printed symbol is called *symbology*. Two symbologies may use the same or different symbol elements to encode the same character string. Some characteristics of a symbology are as follows:

- **Encoding technique.** A symbology with better encoding techniques allows for efficient and error-free encoding.

- **Character density.** A symbology that offers better character density can represent more textual information per unit physical area.

- **Error-checking techniques.** A symbology with better error-checking capability can allow the data to be read correctly even if some of the symbol components are damaged or missing.

About 270 different symbologies have been invented to support specific requirements, and approximately 50 are in widespread use today. Each symbology falls into one of the following three categories:

- **Linear.** Linear symbologies consist of vertical lines with different widths with white space separating two adjacent lines. The maximum number of characters that can be encoded with a linear symbology is up to 50.

- **Two dimensional.** Two-dimensional symbologies have the most data-storage capacity. The maximum number of characters that can be encoded with a two-dimensional bar code symbology is 3,750.

- **Three dimensional (also called a *bumpy bar code*).** A three-dimensional symbology is actually a linear bar code embossed on a surface. This type of bar code is read using the "bumpiness" or the three-dimensional relief of the bar code. A bumpy bar code is thus *not* dependent on the contrast between the bar code lines and spaces for its reading (discussed in the next section). This type of bar code can be painted and subject to harsh environmental conditions, whereas a paper bar code in similar scenarios is easily destroyed.

Subsequent sections discuss symbologies in more detail. For now, however, the discussion turns to operating principles of bar codes and bar code readers, followed by the advantages and disadvantages of the technology.

6.1.2   How Are Bar Codes Read?

Bar code readers, also called *scanners*, read bar codes. A bar code scanner uses a light beam to scan across the bar code. The direction of scanning, in general, is irrelevant. However, during scanning, the light beam cannot move out of the bar code region. Therefore, in general, an increase in a bar code length also means an increase in scanner height to accommodate for larger deviations of the light beam outside the bar code region during scanning. During the scanning process, the reader measures the intensity of the reflected light by the black and white regions (for example, vertical bars) of this bar code. A dark bar absorbs light, and white space reflects light. An electronic device called a *photodiode* or a *photocell* translates this light pattern into an electric current (or analog signal). Electric circuits then decode this generated electrical current into digital data. This data is what was originally encoded by this bar code. The digital data is represented as ASCII characters. A single bar code reader can read several symbologies. Figure 6-1 shows the process just described.

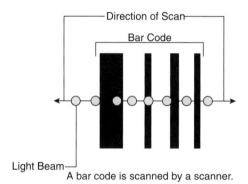

A bar code is scanned by a scanner.

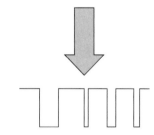

The generated analog signal is digitized.

7   9   5   8

The digitized output is decoded to get
the bar code data.

Figure 6-1   Steps in reading a bar code.

6.1.3   Bar Code Readers

Today, the following four types of bar code readers are available:

- **Pen readers (also called *wands* or *contact wands*).** This type of reader looks like a wand or a pen with the light source focused at the tip. The bar code needs to be in contact with the reader at all times while being read. The advantage of this reader is that it has no moving parts; the user does the scanning manually. As a result, these readers are inexpensive and lightweight (besides being rugged). One of the drawbacks of this type of reader is when a bar code is placed on a roughly textured object. If the bar code's surface is not sufficiently flat, a reader might not be able to read the data correctly. Figure 6-2 shows an example wand reader.

Figure 6-2   A wand bar code reader from Intermec Corporation.

Reprinted with permission from Intermec Technologies Corporation

- **Laser readers.** This type of reader is the most frequently used bar code reader. A laser beam located inside the reader automatically scans a bar code. One of the advantages of this reader type is its capability to read a bar code even if the bar code surface is not flat. It can do so because laser beams of such readers can be focused precisely into a small beam. The laser beam of a reader of this type can either move automatically or be stationary. The laser beam in an self-scanning reader moves back and forth rapidly at the rate of 40 to 800 times per second. Generally, just one scan is needed to read a bar code.

Therefore, this type of reader can read bar codes at a high rate, even if a bar code is of bad quality. A static beam reader is frequently used in industrial operations where the object that has the bar code is moving at a constant speed (for example, on a conveyor belt). In this case, there is no need for a moving scanning beam. A reader of this type can be either stationary or handheld. The maximum reading distance for this type of reader is about 30 feet (9 meters approximately). Figure 6-3 shows an example laser reader.

Figure 6-3   A laser bar code reader from Intermec Corporation.

Reprinted with permission from Intermec Technologies Corporation

- **Charged coupled device (CCD) readers.** This type of reader can read a bar code in a contactless manner. An array of several hundred small light sensors is located at the front of the reader. When the image of a bar code is projected into these photo-detectors, they generate a voltage pattern. This pattern is identical to the voltage pattern generated by a laser reader for this bar code. Some of these systems use additional sources of light, such as a flash, to increase the focal distance. The maximum reading distance for this type of reader is 6 inches. One of the drawbacks of these types of readers is that they cannot read long bar codes because of their limited field of view. In addition, the number of photo sensors in a reader determines the bar code density it can read. Figure 6-4 shows an example of a CCD reader.

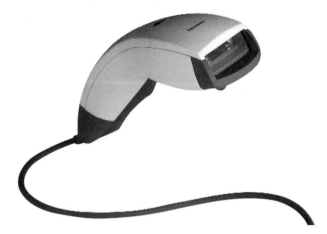

Figure 6-4   A CCD bar code reader from Intermec Corporation.

Reprinted with permission from Intermec Technologies Corporation

- **Camera readers.** These readers are the result of recent advances in bar code technology. A small camera inside this reader captures the bar code image. This image is then
  processed using digital image-processing technology to determine the bar code data. A
  drawback of this reader type is that it is sensitive about the quality of the bar code for an
  accurate reading. For example, the bar code must have sufficient contrast between its
  white and dark symbols and cannot have spots or empty spaces. Camera-based imaging
  scanners have become smaller, faster, and cheaper. A large number of end users are
  replacing laser scanners with imagers for two-dimensional bar code applications. Figure
  6-5 shows an example of a camera reader.

Figure 6-5   A camera bar code reader from Intermec Corporation.

Reprinted with permission from Intermec Technologies Corporation

6.1.4   Benefits

The major benefits of bar codes include the following:

- **Rapid and accurate data collection.** A bar code automates data collection. Using laser readers, you can scan several bar codes in a short period of time. Bar code reading is accurate, with an average error rate of one in three million reads. Table 6-1 (shown later in this chapter) shows a bar code accuracy survey result.

- **Increased operations efficiency.** The data decoded by a bar code reader can be fed directly to an application running on a computer system. Therefore, you can automate various operations, such as price retrieval, personnel identification (e.g., library member), inventory monitoring and control, and so on.

- **Reduced operations cost.** Bar code use offers cost savings by reducing the data-collection errors, reducing manual labor cost, and eliminating process inefficiencies.

6.1.5   Drawbacks

The major drawbacks of bar codes include the following:

- **Easily damaged.** A bar code can be damaged by dirt, paint, fading due to bright sunlight, and moisture.

- **Reader (contactless) operations can be affected by moisture in the environment.** Light rays from a reader are refracted by suspended water particles in the environment, resulting in distortion of focus. Therefore, this reader might experience a loss in read accuracy.

- **Presence of obstacles.** A bar code reader needs to have a line of sight to the bar code it is supposed to read. Any obstacle between the reader and the bar code will prevent reading of this bar code.

- **Speed.** A bar code reader might not be able to read every bar code if they move at a high speed (for example, when the scan rate of a reader is exceeded by the movement speed of the bar codes).

Now, some major bar code symbologies are discussed.

6.1.6   Example Symbologies

It is beyond the scope of this book to explain every symbology that exists today. Therefore, this book discusses only a few of the major symbologies in widespread use today, as follows:

- **Linear.** The following symbologies are discussed for this symbology type:
  UPC
  EAN
  Code 128

- **Two dimensional.** The following symbologies are discussed for this symbology type:
  PDF417
  Aztec code
  DataMatrix

6.1.6.1   UPC

UPC stands for *Uniform Product Code* and is managed by the *Uniform Code Council* (UCC). Two major UPC types are as follows:

- **UPC-A.** This symbology consists of 12 digits, of which the last digit is used as a check digit. The first digit represents the product type, the next five digits the manufacturer code, and the subsequent five digits identifies the actual product. This symbology is used extensively in retail.

- **UPC-E.** This consists of seven digits, of which one is used as a check digit. UPC-E is also sometimes called *zero suppressed UPC* because it can compress a UPC-A code into a six-digit code by suppressing its trailing zeros for the manufacturer code and leading zeros of the actual product. The seventh digit is used as a check digit for the first six digits. Thus, UPC-E can always be converted back into UPC-A. This symbology is used for small retail items.

A supplemental two- or five-digit number can be appended to both UPC-A and UPC-E. Periodicals and publications use this supplemental. Figures 6-6 and 6-7 show example UPC-A with a two- and five-digit supplemental, respectively. Figure 6-8 shows an example UPC-E bar code.

Figure 6-6   An example UPC-A bar code with a two-digit supplemental.

Figure 6-7   An example UPC-A bar code with a five-digit supplemental.

12345678901

Figure 6-8   An example UPC-E bar code.

6.1.6.2 EAN

EAN stands for the *European Article Numbering* system, which is the European version of UPC. Two major EAN types are as follows:

- **EAN-13.** This symbology is the European equivalent of UPC-A. Compared to UPC-A, an EAN-13 symbology contains an additional digit, which together with the twelfth digit generally represents the country code. This symbology is used by the publishing industry to represent ISBN numbers for books. An ISBN code is an EAN-13 bar code with the first three digits as 978 and the remaining nine digits representing the first nine digits of the actual ISBN number.

- **EAN-8.** This consists of eight digits, of which the first two are used for the country code. The next five digits are used for data, and the last one is used as a check digit.

A supplemental two- or five-digit number can be appended to both EAN-13 and EAN-8. Periodicals and publications use this supplemental. For ISBN codes, this supplemental number starts with 5, and the remaining four digits are used to encode the price of the book. Figures 6-9 and 6-10 show example EAN-8 and EAN-13 bar codes, respectively.

12345678

Figure 6-9   An example EAN-8 bar code.

1   234567   890123

Figure 6-10   An example EAN-13 bar code.

6.1.6.3   Code 128

This variable-length symbology uses both alphabetic symbols and digits. This symbology is widely used and is generally considered to be the optimal choice for a variety of bar code applications. It uses characters from the following three sets:

- Uppercase alphabets and ASCII control characters
- Uppercase and lowercase alphabets
- Numbers from 00 through 99

The first character is a special character that signals which of these sets is used initially. Three special shift codes, one for each set, also allow changing the character set after the initial set. This symbology also uses a check digit. Figure 6-11 shows an example Code 128 bar code.

a1b2c3

Figure 6-11   An example Code 128 bar code.

6.1.6.4   PDF417

PDF stands for *Portable Data Format*. This high-density, two-dimensional symbology, invented by Symbol Technologies, Inc., can encode all 256 ASCII characters. A maximum of 2,525 characters can be represented by a bar code of this type. This symbology consists of smaller bar codes stacked on top of each other. This is a mature symbology that provides several options such as data security and compression, error detection and correction, and so forth. This bar code offers the next-best read accuracy (see Table 6-1) after DataMatrix. Figure 6-12 shows an example PDF417 bar code.

Figure 6-12   An example PDF417 bar code.

6.1.6.5 Aztec Code

This high-density symbology can encode all 256 ASCII characters. A maximum of 3,750 characters can be encoded by a bar code of this type (when the characters are all digits). The basic building blocks of this symbology are shaped like a square and are called *modules*. At the center of this bar code is a square-shaped bulls-eye surrounded by layers of encoded data. A bar code of this type can be read independent of its orientation. Figure 6-13 shows an example Aztec bar code.

Figure 6-13   An example Aztec bar code.

6.1.6.6 DataMatrix

This high-density symbology can encode all 256 ASCII characters. A maximum of 3,116 characters can be encoded by a bar code of this type. A distinguishing characteristic of this symbology is its perimeter pattern. The new version of this symbology, called ECC200, offers better encoding and better error-detection and -correction schemes. This bar code offers the maximum read accuracy (see Table 6-1). Figure 6-14 shows an example DataMatrix bar code.

Figure 6-14   An example DataMatrix bar code.

6.2 Advantages of RFID Over Bar Codes

The advantages of RFID over bar codes are as follows:

- **Support for nonstatic data.** An RFID tag data can be rewritten many times (assuming, of course, that the RFID tag is an RW tag). The data on a bar code is static and cannot be changed.

- **No need for line of sight.** Generally, an RFID reader does not need a line of sight to read an RFID tag's data. A bar code reader *always* needs a line of sight to read a bar code.

- **Longer read range.** An RFID tag can have a much longer read range than a bar code. Depending on several factors, this can range from several feet to a few hundred feet

- **Larger data capacity.** An RFID tag can store more data than a bar code.

- **Multiple reads.** A suitable reader can read several RFID tags within a very short period of time, automatically, using a feature called *anti-collision*. A bar code reader, however, can only scan one bar code at a time.

- **Sustainability.** An RFID tag is generally rugged and resistant to harsh environmental operating conditions (to a fair extent). A bar code is easily damaged (for example, by moisture or dirt).

- **Intelligent behavior.** An RFID tag can be used to do other tasks besides simply being a data carrier and transporter. A bar code, however, does not have any intelligence and is a vehicle for only storing data.

The following, although often mentioned in the media, are dubious and therefore are not considered *clear* advantages of RFID over bar codes:

- **Read accuracy.** RFID is far more accurate than bar codes.

- **Item-level tagging.** A bar code does not support item-level tagging.

The following sections discuss these points in more detail.

6.2.1   Support for Nonstatic Data

You can rewrite the data on a read-write RFID tag many times. How many times? In general, you can rewrite an RW RFID tag on the market today at least 10,000 times, with vendors claiming 100,000 times or more! Rewriting proves useful if you use the tag to record data that was not available when it was *created*. A tag is created when some data is initially put on the tag to make the tag usable. This data might be an identifier that identifies this tag uniquely from the set of all possible tags. Other types of additional information (generally about the object on which this tag is attached) are also possible. Why would you want to rewrite the data on a tag? It depends on the application for which this tag is used. For example, suppose that an RW tag is attached to an item that is being built as it moves through a production line. As the item moves through the different stages of the production line, the time taken to complete each stage can be recorded on the tag. Finally, when the item rolls off the production line, the data recorded on the tag can be used to analyze, for example, the production-line bottlenecks (that is, places where this item is spending most of its time). Also, if an RFID tag needs to be recycled in an application, it may need to rewrite its old data with new data.

A bar code, in contrast, can only store static data (that is, data that cannot be rewritten with new data). A new bar code has to be created every time to store this new data.

6.2.2   No Need for Line of Sight

One of the distinct advantages of RFID is that it does not require line of sight. Suppose, for instance, that an RFID tag is attached to an object and that object is placed inside a container made of some RF-lucent material (for the frequency used) such as paper. An RFID reader can read this tag right through the container without any need to open it! Therefore, the reader does not need to "see" the tag to read its data. In some situations, however, this is *not* true. In these situations, a line of sight *must be* established between the reader and the tag for the tag data to be read reliably. (Although establishing a line of sight does not guarantee improved readability, it helps to configure the critical factors such as tag read distance, reader energy, and reader antenna positioning to counter the environmental impact.) In these situations, UHF tags are used, *and* there is a significant amount of RF-reflecting material, such as metal, present in the operating environment. A classic example is an automobile production line, where essentially everything is made of metal and there is a high degree of RF wave reflections in the surroundings. If a UHF RFID tag is placed on a vehicle to track it in the production line, the tag and the readers must be placed in such a manner that a line of sight can be clearly established at the points where readings will take place (to avoid multipath signals and antenna nulls). In these situations, an RFID tag offers no advantage over bar codes.

A bar code reader always needs a clear line of sight to a bar code to scan it. Therefore, it might perform poorly in some typical applications, such as airline baggage handling (where the tagged objects [baggage] are oriented randomly on the luggage belt). In this scenario, a good chance exists that overlapping bags will obstruct the line of sight to a bar code on other bags. In addition, a bar code on a bag might be oriented in such a manner that the bar code reader cannot read it. These factors will lead to poor read performance. If suitable RFID tags and readers are used, however, orientation of the tag to the reader might not have a significant impact as compared to bar codes. In addition, because line of sight is not needed, most of these tags can be read through overlapping suitcases (assuming made of RF-luent material for the frequency used). Therefore, read accuracy of RFID luggage tags can be substantial higher (more than 15 percent higher has been reported) than with bar coded tags.

6.2.3   Longer Read Range

A passive RFID tag operating in the UHF range has a read range of about 30 feet (about 9 meters) under ideal conditions. An active tag in the low UHF range (433 MHz) has a read range of about 300 feet (about 91 meters). An active tag operating in the 2.45 GHz range has a read range of more than 100 feet (about 30.5 meters).

The principle of bar code reading is tied to optics, and the read range of a bar code reader depends on the focal range of the reader. The read range of commercially available readers is 30 feet (about 9 meters) or less.

6.2.4   Larger Data Capacity

A two-dimensional bar code, such as Aztec, can encode up to 3,750 characters of data from the entire 256 ASCII characters, which is substantial. There is not a huge difference today in data capacity as far as passive tags are concerned, but this might change in the future. Custom active tags, theoretically, have unlimited data capacity.

What does a tag with large memory buy for its user? This is a loaded question for which there is no easy answer. Clearly, for an active tag that needs a large amount of memory to perform specialized tasks, a large tag memory is an advantage. With regard to the amount of data transferred, the transmission time and the error rate of reading increases with an increase in data transfer size. Therefore, a tag with a large memory capacity for (storing and) transferring a large amount of data to a reader can prove to be a disadvantage!

6.2.5   Support for Multiple Reads

An RFID reader can automatically identify a few to several tags in its read zone in a short period of time. This capability to automatically identify multiple tags can make tag-reading operations fast.

A bar code reader can only read one bar code at a time, which means a longer read time of bar codes compared to the same number of RFID tags by an RFID reader. Therefore, if the cost of an RFID system is acceptable, existing operations that use bar codes might be made more efficient.

6.2.6   Sustainability

An RFID tag can (to a fair degree) withstand harsh environmental conditions such as heat, humidity, corrosive chemicals, mechanical vibration, and shock. Note that currently *no* single tag can withstand *all* these environmental conditions. Generally, a particular tag from a specific vendor is resistant to one or a few of these conditions.

A bar code is only as good as the material on which it is printed. Therefore, a bar code printed on paper is easily damaged in the presence of moisture or heat. A bar code can be easily smudged in presence of dirt and paint. A bumpy bar code can withstand high temperatures.

6.2.7   Support Intelligent Behavior

An RFID active tag can carry on-board electronics and a power supply (battery) to perform functions such as monitoring its surrounding temperature, humidity, and so forth. The tag can then use this data to dynamically calculate other parameters and transmit it (with its unique ID) to a suitable reader.

In contrast, a bar code is just a repository of static data and nothing else.

6.2.8   Read Accuracy

Now back to the original issue of read accuracy. A common statement today is "RFID is far more accurate than bar codes." However, this statement has two problems. First is the *availability* of official, objective data on the accuracy of bar codes. Second is the *complete lack of availability* of official, objective data on the accuracy of RFID tag reads over bar codes. With an absence of hard data on RFID accuracy, but the presence of hard data about bar code accuracy, it is unfair to say RFID is "far more accurate" than bar code.

In the classic bar code accuracy study conducted by Ohio University, using DataMatrix symbology, the worst rate of error was 1 per 10.5 million; the best rate of error was 1 per 612.9 million reads! These are *accurate* numbers indeed! Table 6-1 lists a summary of findings.

Table 6-1   Bar Code Accuracy (Summary of Ohio University's Findings)

Symbology	Worst Case	Best Case
DataMatrix	1 error in 10.5 million	1 error in 612.9 million
PDF417	1 error in 10.5 million	1 error in 612.4 million
Code 128	1 error in 2.8 million	1 error in 37 million
Code 39	1 error in 1.7 million	1 error in 4.5 million
UPC	1 error in 394,000	1 error in 800,000

From my experience with clients who have existing automated bar code systems in production, the accuracy of bar code reads (after system tuning) is typically in the 90 percent range or higher. Therefore, the accuracy advantage of RFID over existing bar code systems seems equal or less than 10 percent (in general, and at the best) for these types of bar code solutions. In some situations, this can hardly be called "substantial," assuming, of course, that RFID will actually increase the accuracy rate at this level. If the environment is not well suited for RFID, the accuracy improvement rate using RFID over bar code might be zero (or even negative)! Then again, depending on the application, RFID might offer accuracy benefits of more than 10 percent (the airline baggage-handling example, for instance). In addition, for a particular application, an increase of even a few percentage points might bring substantial value to a business.

6.2.9   Support Item-Level Tagging

There seems to be a growing belief that only RFID can support item-level tagging, whereas bar code cannot. This is *untrue*. Different types of bar codes have varying capacity for storing data. The linear bar code types that are used most widely for item level tagging (for example, UPC) do not have sufficient storage to identify an item uniquely. However, other bar code types, if used, have more than enough characters to identify *any item uniquely*. For example, assuming alphanumeric data storage, DataMatrix can store up to 3,116 bytes; Aztec 3,750 bytes; and PDF417 1,850 bytes. These capacities are more than enough to contain a unique number of 1,024 bits, which itself is more than enough to uniquely identify any particular item.

At this point, you might thinking that, with so many advantages, RFID is a clear winner over bar codes. Hold this thought. Although RFID has several advantages over bar code, the next section covers the advantages bar code has over RFID.

6.3   Advantages of Bar Codes Over RFID

The advantages of bar codes over RFID are as follows:

- **Lower cost.** The cost of implementing a bar code solution is generally less than that of a comparable RFID solution.

- **Comparable accuracy rates.** In several cases, the accuracy of a bar code solution is about the same, if not better, compared to an equivalent RFID solution.

- **Unaffected by the material type.** A bar code system can be used to successfully tag almost every kind of material.

- **Absence of international restrictions.** Bar code systems are used worldwide without any legal limitation on the use of the technology.

- **No social issues.** Today, you can find bar codes on almost every item on the planet, but no privacy rights group object to its use.

- **Mature technology with large installed base.** Bar code technology is probably the most widely deployed technology in the world.

The following sections discuss these points in detail.

6.3.1   Lower Cost

The cost of a bar code is close to zero, whereas cost of an RFID tag is about 20¢ or more for UHF tags when ordered in large quantities. In addition, an average price of a bar code handheld reader is less than $400; the cost of an RFID handheld reader is more than $800 for UHF readers. Similarly, stationary bar code readers on an average cost less than $700, whereas RFID readers currently cost more than $900 (UHF readers). For 13.56 MHz readers, the cost of handhelds and stationary readers is generally less than the UHF readers, but 13.56 MHz tags are generally more expensive than UHF tags. For RFID, additional hardware is required—antennas are also needed besides the reader. The cost of an RFID antenna (both linear and circular) typically ranges from $150 to $500. More expensive antennas also exist. The cost of an RFID reader and the antennas (generally two per reader) drives the price differential between the bar code and RFID readers even higher. On the other hand, a high-quality, long-range (about 15 to 20 feet) stationary bar code reader might cost several thousand (more than $5,000 is not uncommon). For high-quality handheld bar code readers, the cost could be more than $4,000. These prices are more than most of the RFID readers available on the market even when the typical cost of antennas (two antennas per reader) is factored in. However, these expensive bar code readers might not be necessary for a typical application. Even assuming the difference in cost between the highest-priced bar code readers and the cheapest RFID readers and antennas, the recurring cost of RFID tags (assuming that the tags are not recycled) will ultimately overrun any cost savings thus achieved.

6.3.2 Comparable Accuracy Rates

The current bar code systems installed in production systems have generally high read accuracy. Read accuracy in the range of 90 percent is common, and accuracy of 98 percent is not uncommon. Therefore, for these types of applications, RFID cannot offer more than a 10 percent increase in accuracy, assuming, of course, an equivalent RFID system will actually work better. Depending on the operations environment and other factors, such as tagged object material type and content, an RFID solution might actually do worse than the equivalent bar code solution.

6.3.3 Unaffected by the Material Type

A bar code can be put on an object made of almost any material, regardless of whether it is RF-lucent or RF-opaque for the RFID frequency used. RFID tags can be read with difficulty, if at all, on metal and some liquids in UHF and microwave frequency ranges. Therefore, if an environment has too much metal in it, an RFID system might not work well when operating in these frequencies.

6.3.4 Absence of International Legal Restrictions

Bar code technology works on optics principles, whereas RFID technology works on the principle of RF waves. This distinction has an important bearing on the legal limits of use of the technology. No international limit applies to frequency of light, but quite a few restrictions apply to RF waves. Widely varying international limits apply to RFID system frequency ranges and with reader transmission power. Therefore, an RFID system built for a particular frequency type in one country might not be legally compatible in another country or might require nontrivial modifications that result in multiple systems for essentially the same application. As RFID evolves and its acceptance increases, some of these restrictions might disappear as governments cooperate to loosen the frequency and power restrictions to reap the benefits of RFID. Multiple-frequency readers from vendors offer an alternative solution to this issue.

6.3.5 No Social Issues

A bar code has no social issues tied to its use because the bar code type that is put on an item is only meant to identify the product type and provide other information such as price in a generic manner. Thus, a bar code on a packet of potato chips identifies the packet as containing potato chips and the price. However, it does not identify a bag of potato chips uniquely from another bag of similar potato chips. Not that this cannot be done using a bar code type that stores more data, but it is not done today. This anonymity obviates social issues, such as privacy-rights infringement concerns, that currently impact the acceptance of RFID. Tagging an item with a bar code is accepted today worldwide without raising any eyebrows, but trial efforts using RFID have caused public uproars from privacy-rights groups and lawmakers. Until legal, business, and technological interests settle this debate, it might present a hurdle to ubiquitous acceptance of RFID (see Chapter 5, "Privacy Concerns").

6.3.6   Mature Technology with Large Installed Base

Bar code has been in existence for the past 30 years. In these years, the technology has matured tremendously. More than 50 bar code standards are currently in widespread use today, and several of these standards (for example, UPC and EAN) enjoy widespread support around the world. Today, bar codes are ubiquitous in every facet of the economy. Currently, it is estimated that every day, about five billion bar codes are scanned. In fact, bar codes are so ubiquitous today that they seem to be mundane and hardly worth consumer notice—a true sign of a successful technology.

Compared to the success of bar codes, the success of RFID today can be said to be extremely limited both in terms of the array of current application types and the collection of application members inside such types. RFID technology is considered an emerging technology, and as such, it is still in its infancy. The price and performance of the hardware, the complexity involved in designing a medium to large solution, and the privacy-rights infringement concerns for item-level tagging might delay RFID widespread adoption for some time to come.

RFID and bar code technology taken together do not cover every application that is possible. Both have common disadvantages, which are covered in the following section.

6.4   Disadvantages of RFID and Bar Codes

The major disadvantages of both RFID and bar codes are as follows:

- **Presence of obstacles.** A bar code reader cannot read a bar code if there is any obstacle between the reader and the bar code. An RFID reader, depending on its operating frequencies and other factors, such as power and duty cycle, might not be able to read a tag if there are any RF-opaque obstacles, such as metal, or RF-absorbent material, such as water present between the reader and the tag.

- **Presence of moisture.** For bar code readers, the light beam might be refracted by water particles suspended in the atmosphere, resulting in focus distortion. For RFID readers operating in UHF and microwave frequencies, water particles suspended in the atmosphere might absorb RF energy, resulting in insufficient energy reaching the tags for proper data transfer.

- **Speed.** If the scan rate of a reader is exceeded by the speed of movement of the bar codes, a loss of reading accuracy, together with failure to read a bar code, might result. For RFID readers, if the speed of the tag is so great that the tag has insufficient time to optionally energize itself properly and transmit data back to the reader, a loss of reading accuracy, together with failure to read a tag, might result.

- **Extrinsic identification schemes.** A bar code or an RFID has to be applied externally to an object; these are not part of the object's physical characteristics. Therefore, if such an object is mislabeled or "mis-tagged," the object identity is in jeopardy. However, perhaps intrinsic properties could be used to uniquely identify an object. For example, a fingerprint or a retinal scan of a person can uniquely identify this person without any need to put an external identification scheme, such as a bar code or RFID tag, on the person.

You are now ready to explore what can be called the most-hyped potential of RFID in the media.

6.5   RFID Will Replace Bar Codes Soon

The short advice is this: "Forget it." Read on to understand why. For RFID to "replace bar codes soon," it must overcome the following hurdles "soon":

- **Tag any item that a bar code can tag today.** Such items include almost every type of physical merchandise in existence in the world economy. To do this at an acceptable cost, the following four hurdles must be overcome:

  - *Cheap hardware with tags costing less than 5¢.* The profit margin of some of industries is razor thin and prone to cutthroat competition. Any extra cost that does not go toward the bottom line is rarely justified.

  - *No consumer issues.* The consumer must accept the use of RFID to tag every item that bar codes can today.

  - *Technical advancement to satisfactorily tag any possible item.* RFID is an emerging technology, so the capabilities of tags, readers, and antennas are all undergoing rapid changes. At this point, the capabilities are not sufficient to tag every item to which a bar code can be affixed.

  - *Worldwide acceptance of common frequencies of operation.* When common frequency bands for RFID operations are standardized, deployment of RFID implementations will definitely speed up. Even if these hurdles are overcome, there is still the following last hurdle, which might be the most daunting of all.

- **Replace a tremendously large base of working bar code solutions.** Even if RFID resolves all the preceding issues to become at par with bar codes, why should a business invest money to replace a perfectly acceptable and working bar code solution?

These hurdles are now discussed in detail in the following sections.

6.5.1   Tag Any Item That a Bar Code Can Tag Today

Clearly, if RFID is to dislodge bar code from its current status as "uber tag," it needs to tag any item that a bar code can today. This issue has three main components: economic, technical, and social. Each of these three components must be resolved before RFID can level the playing field with bar code. The next two subsections discuss the economic and social components, respectively, followed by a technical component discussion in the following two subsections.

6.5.1.1   Cheap Hardware with Tags Costing Less Than 5¢

To successfully par with bar codes, RFID tags need to be cost effective. The cost of producing a bar code is next to free, but the cost of an RFID tag is very expensive today when it comes to tagging low value items for which the profit margins are low. Manufacturers and retailers want to maximize their profits and any additional overhead that does not help their bottom line are almost always ruled out. Thus, for an item for which the profit margin is 10¢ it does not make sense to put a 20¢ UHF RFID tag on it, which is what an average UHF RFID tag approximately costs when purchased in large volume. As a result, the tags have to be as cheap as possible. There seems to be common consensus today that for tagging individual consumer items, the price of a tag has to fall below 5¢ and for several other items, the price has to go under 1¢. At the same time, the price of RFID readers and antennas have to come down to below $100. These are tall orders considering the manufacturing process and cost of producing current RFID hardware. For example, in order to produce an RFID tag, several manufacturing steps are necessary, each of which adds a fixed overhead to the cost of manufacturing and hence to the final cost of the tag (see Appendix B, "Manufacturing Overview"). The tag cost can be lowered in one of two ways. The first option is to eliminate some of the processing steps, resulting in more defective (but cheaper) tags (which might or might not be acceptable to the business). The second option is to invent new manufacturing processes that consolidate a number of the steps, thus bringing down the cost while maintaining the tag quality or even improving it. This second option seems to hold the most promise. When will 5¢ and 1¢ tags appear on the market? The industry speculates that such tags are about 5 to 10 years away from being available commercially.

Tag costs do not make exciting business models for manufacturers. To build a $200 million fabrication plant making 1 billion tags at 1¢ each, for example, it will take 20 years to realize *return on investment* (ROI), assuming 0 inflation! When will sub-$100 readers and antennas appear on the market? There is no current estimate on the readers and antennas. Although rapid advances are being made in these areas, it seems safe to speculate that these are at least 3 to 5 years away from being available commercially. Again, however, the price of the tags also must come down to an acceptable level, not just the readers and antennas. Therefore, it seems that RFID still has about 5 to 10 years to be at par with bar codes in terms of cost.

6.5.1.2   No Consumer Issues

The current ongoing debates, protests of privacy-rights groups, and attempts by legislators to impose legal regulations seem just starting to take shape with regard to the use of RFID to tag individual consumer items. A consensus between the technology backers and the privacy-rights supporters does not seem near. It could be years before any conclusive decisions are reached; RFID tagging of consumer items might take a back seat until then. Therefore, even though the technology might provide sufficiently cheap tags to tag individual items five years from now, this fact alone does not imply that RFID item-level tagging will be actually put into practice at that point.

Because of the regulations and standards that might apply to RFID in the future, RFID might be used just to tag certain types of items, with bar codes used to tag the remaining items. RFID then loses its battle with bar code right there.

6.5.1.3   Technical Advancement to Satisfactorily Tag Any Possible Item

RFID technology is still in its infancy when it comes to tagging different item types. The RF properties of an item, its physical characteristics (such as shape), operations environments, and so forth all have important bearings on tag size and properties. A wide array of tags with various characteristics might be needed to satisfy these requirements. This field is experiencing fast improvements, but even so, tag technology is far from reaching a mature level (which might, in fact, take another 10 years or more).

However, bar code technology and its adoption rate is not going to stand still for the next 5 to 10 years. For example, the UCC (the overseer of the most popular bar code, UPC) has mandated through its 2005 Sunrise program in 1997 that by January 1, 2005, all U.S. and Canadian businesses must be able to scan and process EAN-8 and EAN-13, besides 12-digit UPC symbols, at the point of sale. This program is aimed at promoting global commerce and facilitating commerce efficiency. While not every company in the U.S. and Canada complied with this mandate, it shows that the bar code industry is quite active and innovative.

6.5.1.4 Worldwide Acceptance of Common Frequencies of Operation

A common worldwide standard of RFID frequency (UHF in particular) will act as an accelerator of acceptance of the technology. With this, a single RFID system implemented for a particular business application can be deployed worldwide without any costly changes to suit country-specific regulations. This standardization will reduce the maintenance efforts of the solutions and might allow the solutions to be standardized for a particular application. RFID solutions, which can then be bought virtually off the shelf and put into use in any part of the world, will act as a strong catalyst for acceptance. Note that to bypass this issue, vendors are bringing out readers that can operate on multiple frequencies, which might offer a solution to this issue.

However, such a common, worldwide acceptance, assuming it happens, is years away. Ten years could be insufficient for this purpose because the technology has to be mature enough in the first place to convince world governments that it makes sense to invest their resources on an effort.

6.5.2 Replace a Tremendously Large Base of Working Bar Code Solutions

So far, this chapter has discussed how RFID technology can overcome its acceptance hurdles. Assuming that RFID does just that, however, which is an extremely big assumption, does it guarantee a replacement of bar codes? Well, not really! The reason is chiefly economical—what are the business drivers to shut down perfectly working bar code solutions used by an extremely wide range of large, medium, and small businesses and replace these with RFID? When the item moves from large manufacturers to retailers to small businesses, the chances of gaining huge productivity gains seem far away, because the volume of handled goods also decreases substantially (resulting in a tighter room for improvement). In addition, the cost of implementation and maintenance of an RFID system has to be less than the maintenance cost of the existing bar code systems unless strong business drivers exist to offset the cost difference with productivity gains. That RFID can do this for every existing bar code system in the world is difficult, if not impossible, to anticipate.

From the analysis in this chapter, you can understand that there is little possibility of RFID replacing bar codes completely in the long run, let alone "soon." Both RFID and bar codes will coexist as complementary technologies for years to come. This conclusion might lead you to think that RFID is not a useful technology after all. Don't despair yet, however; the next section will convince you otherwise.

6.6   Conclusion

A comparison to bar codes did not make RFID look as promising as the current hype makes it out to be. In fact, you might be wondering whether RFID is heading toward failure because it cannot replace bar codes. *This is a completely incorrect opinion*! Although it might be true that RFID will not replace bar codes, such a fact does not herald the demise of RFID technology. Why? *Because there is life beyond bar codes*! No law states that RFID must replace bar codes to be successful. An untold number of applications for which bar codes are totally out of context represent a wide opening for RFID. Examples of these applications include smart tags, anti-tampering tags, and tags to locate objects in real time; for these scenarios, RFID might just be the right technology.

The problem with the bar code versus RFID debate does not have to do with the technology, but with the misguided zeal to prove the superiority of RFID. Hype that one of the most prevalent technologies today will be defeated "soon" attempts to make people notice and accept RFID. As it turns out, however, such hype targets a competing technology that is hard to defeat with the current state of RFID technology.

The comparison of the two technologies in this chapter exposed some of the major improvements RFID technology needs. Therefore, two things might have happened in this chapter: First, you might have seen through the hype, and this understanding might have created a negative opinion about the technology; second, the current hurdles RFID faces might have prompted you to create some hype of your own regarding the "failure" of the technology. Please just let the facts guide your opinions.

Whether RFID can replace bar codes does not prove much about the success or demise of the technology. Bar code applications are just some of the areas where RFID can be applied. In fact, RFID can be applied in applications that lie completely beyond the reach of bar codes. Therefore, these two technologies have distinct capabilities and ranges of applications where they perform well. Any comparison that leads to the demeaning of one technology over the other is not rational. As RFID matures, the technology might induce the development of applications that are currently thought too difficult or even impossible. Finally, note that many considered bar code systems a failure at their inception during the 1970s. Since then, bar code technology has come a long way (to probably the most used technology in the world today). From this perspective, it is safe to write that RFID has the chance to become one of the most used technologies in the world 30 years from now.

CHAPTER 7

The RFID Strategy

An RFID strategy provides a roadmap to use the technology aligned with an enterprise's strategic vision and goals. For example, a business that strives to be a model of efficiency could use RFID to streamline its operations. An RFID strategy is strongly recommended for a large enterprise. A smaller scale company may also benefit from such a strategy. An RFID strategy also shows the extent to which a business is ready to use RFID within itself.

A one-size-fits-all strategy is generally not possible, which means that businesses must create their own unique RFID strategy, determine how RFID can create value that is aligned with its strategic directions, factor in such external drivers as meeting customer RFID mandates all within the tolerable cost/risk ranges, and so on. The following section examines the benefits of having such a strategy in place. This chapter then provides some high-level example strategy guidelines.

7.1 Why an RFID Strategy?

Do *not* mistake the creation of an RFID strategy (in the context of a large enterprise) for a needless corporate process that you can short-circuit to delve right into the excitement of implementation. Such a short-circuited implementation, although perhaps successful in the eyes of the implementers, might come to be viewed as a failure, or even useless, by other parts of the business in the long run. Post-implementation use of the technology might be questioned/resisted, leading to frustration and demoralization of the technology supporters. You can avoid these types of situations by ensuring top-down buy-in, and a comprehensive RFID strategy can facilitate such top-level decision making.

In brief, the fundamental reasons for establishing an RFID strategy include the following:

- To determine the various impacts of RFID technology
- To ensure basic understanding and buy-in from senior management

- To form a high-level master plan from which business justification, deployment strategies, and other policies can derive
- To ensure cross-functional support

The following sections review each of these reasons in detail.

7.1.1   Determine the Various Impacts of RFID Technology

RFID is a tool that businesses must apply judiciously to realize its benefits. Otherwise, the use of the technology might turn out to be detrimental. For example, operational efficiencies might go down; costs might rise, resulting in lost revenue, missed opportunities, and customer turnover. Therefore, businesses gain by *not* using RFID technology before determining how best to use it and how that use will impact their processes and personnel. When decision makers do decide to deploy RFID, a comprehensive strategy enables them to validate its use via justification analysis and pilot implementation results. Thus, enterprises can focus on areas that align properly with their line of business and that promise the maximum return on investment while significantly eliminating misdirected efforts that might negatively impact the business.

7.1.2   Ensure Basic Understanding

An RFID strategy can show the potential benefits of using the technology at a level that can be viewed, analyzed, and understood by senior decision makers. In general, each decision maker has his or her own special interests and priorities. The viability of a plan in the long run depends on how these people perceive the plan and understand its benefits for them in their individual endeavors. Therefore, an RFID strategy should cater to a broader set of interests rather than seek to satisfy the needs of a few.

The best way to achieve this goal is to create a strategy that aligns with the core competencies and strategic directions of the business instead of focusing on isolated areas of benefit. Such an integrated plan, when validated and supported at the highest level of decision making, can provide a path of realization that is less cluttered with internal politics, funding nightmares, and deployment-decision delays. The implementation of a successful RFID system being as nontrivial as it gets, having these elements under control will enable the adopters to focus on solution delivery, which is by no means a small achievement in a business environment. In addition, an RFID strategy ensures that if the technology meets the benefit expectations, it will be assimilated into the other parts of business in the long run.

7.1.3   Form a High-Level Master Plan

An RFID strategy can drive the next level of activities, such as the following:

- Estimating the technology benefits
- Estimating the cost of implementation
- Creating a deployment strategy

The outcome of some of these activities can provide further evidence to validate the strategy objectively. Indeed, the strategy needs to be validated periodically from the results of the adoption efforts. Such validation can help the proposer bolster his credibility and can strengthen the case for use of RFID in the business.

7.1.4 Ensure Cross-Functional Support

Enterprise concerns are typically multidimensional, involving several functional areas of the enterprise. For example, chronic shrinkage might involve ordering, security, warehouse and transportation logistics, and so on. If RFID is to be used to address such a problem, a decision regarding its use will probably require input from all the involved parties, generally at senior-management levels. Therefore, cross-functional support is a crucial element for the success of an RFID deployment to solve a business issue. This support generally includes resource allocation, identification of the relevant factors, design and implementation of solution components to address the factors, integration of the solution components to provide an integrated RFID solution, process changes, and training. The buy-in of the top management for an RFID strategy that aims to eradicate problems or improve business efficiency can act as a powerful catalyst to jump-start such interdisciplinary support and cooperation.

The following section discusses some high-level RFID strategy guidelines.

7.2 Strategy Guidelines

The following list of strategy guidelines is by no means exhaustive and is provided here for informational purposes only:

- Enhance security
- Increase operational efficiency
- Enable superior branding

Clearly, these strategy guidelines are quite broad. A business can select one of these guidelines and customize it depending on its core competencies and strategic goals. These guidelines are not mutually exclusive; that is, choosing one does not necessarily preclude choosing another. This is especially true of the first guideline on the list. The following sections provide examples of how you can customize these guidelines. Chapter 4, "Application Areas," provides more details about specific RFID applications mentioned in the following examples.

7.2.1 Enhance Security

Security is probably the most widely needed improvement in any business, but often receives inadequate resources and effort, perhaps because security enhancement, in general, does not directly translate into a company's bottom line. Decision makers often defer or completely ignore the issue of security when faced with the needs of other parts of the business. Security also has many aspects; for example, the means of theft can vary widely. Faced with a myriad of possibilities and the cost of encountering these, a company might despair and decide that security is a losing game for the business. RFID can offer a single-technology solution to deal with a wide variety of commodity thefts. These thefts directly, and perhaps substantially, affect a company's bottom line in lost revenue. Consider the following.

Shrinkage represents a significant problem for manufacturers, distributors, and retailers. Annually, billions of dollars of potential revenue are lost due to shrinkage. Part of this burden is passed to customers via higher prices. Existing anti-shrinkage processes are either largely nonexistent or insufficient to catch and prosecute perpetrators. RFID can provide item tracking and management capabilities. Combined with other security measures, such as RFID-enabled access control, RFID item tracking enables a business to know what item has been accessed by whom and when. In addition, RFID can actively resist a theft attempt by activating real-time alarms, locking access doors, and triggering video monitoring of the suspect and the item being pilfered.

Counterfeiting is another security concern that affects many businesses worldwide. Besides financial loss, a counterfeited item might harm its user (for example, counterfeited pharmaceuticals) and might decrease brand loyalty (among other things). RFID can provide a unique electronic signature for an item that can be used to determine the item's authenticity and provide a plethora of other information such as manufacturing date, target destination, and so on. This information can also be used to determine whether the item has been counterfeited.

Business liability is a serious concern and can arise from various causes (for example, mishandling of hazardous materials, the presence of harmful bacteria in edible items such as meat, or faulty merchandise such as automobile tires). When business liability issues arise, it is necessary to trace the specific items involved. RFID enables companies to do so by uniquely identifying only the items associated with a particular incident so that the company can deal with those accordingly. For example, in the case of faulty automobile tires, the manufacturer can selectively recall only the tires that are known to be defective instead of recalling every tire that it sold over a particular period of time.

National security represents an extremely important concern that requires involvement not only from the government but also from the business community. Terrorists can use items such as hazardous materials and shipping containers, and resources such as drinking water, to launch attacks against a country. RFID can provide effective real-time monitoring and alarm capabilities to deter these types of attempts. A business, looking beyond its operations, competitors, and affiliation with the government, should investigate whether it might be exploited in any way to breach national security and should then take appropriate countermeasures. RFID can provide a one-stop shop for implementing a wide variety of these security measures.

7.2.2   Increase Operational Efficiency

Businesses can use RFID to increase efficiency in manufacturing, supply-chain, and other types of operations, such as internal logistics. Note, however, that any given operation is process-centric. Therefore, a technology such as RFID might offer only limited benefits if the process itself is not amenable to change.

In manufacturing, businesses can use RFID to identify production bottlenecks, accurately customize products, and implement quality control measures, to name a few, which can lead to substantial productivity gains and optimization of the manufacturing processes.

RFID is currently receiving a tremendous amount of attention in supply-chain operations. However, as mentioned previously, it is the process that ultimately needs to be modified to derive the maximum benefit from the technology. Therefore, a supply chain driven by product demand in near real time can take several years to build and require the collaboration of manufacturers, distributors, and retailers. RFID can prove a very useful tool to take advantage of this collaboration and the integrated processes that result from it. However, a business does not have to be optimized at this level to derive advantages from RFID. For example, today, RFID can automate shipping and receiving, reduce out of stocks, and eliminate manual labor.

Businesses can use RFID to increase the efficiency of their internal operations. For example, RFID can facilitate check-in and checkout of rental items and eliminate laborious filling out of paper forms and notes by using a single identification tag on an item. This can also eliminate multiple identification tags/labels that may be used today to identify an item.

7.2.3 Enable Superior Branding

In today's highly competitive business environment, companies constantly search for ways to distinguish their brands and products from competitor products. For example, some companies offer better customer service; others provide useful new products and services that set them apart from the rest of the pack. RFID represents another way for businesses to enhance their branding.

Businesses can use RFID to provide customers information on demand. For example, a tagged item, such as a bottle of wine, might provide information about the place and date of production and composition. Types of food that this particular wine complements might comprise part of this information. Thus, the shopper becomes aware of the quality of the product, and the potential to cross-sell is enhanced (for instance, other products such as the foods suggested). In addition, the retailer can share sales information regarding particular wines with distributors and producers so that they can all position themselves to better meet demand.

The application of RFID is almost boundless, constrained only by one's imagination. A business can come up with creative ways to offer new products or services that can provide it with a competitive edge compared to other similar companies in the market. For example, a business can guarantee that products meet quality standards (ISO 2000, government-approved organic standards, and so forth). Customers themselves who are using (or considering using) RFID technology might also be attracted to companies that are likewise deploying RFID, which might result in more business from these like-minded customers.

7.3 RFID Strategy to Deployment Strategy

An enterprise RFID strategy serves as a high-level roadmap for adoption of RFID technology. How does a business navigate this roadmap to a deployment strategy? This section answers this question.

A business-justification team can use the broad outline of an RFID strategy to identify specific areas in which the technology might deliver its benefits in a relatively quick manner (see Chapter 8, "Creating Business Justification for RFID"). The team uses the corresponding business cases and roadmaps to plan the first iteration of the implementation. Only one business case should be selected for this—the most promising (the top-priority business case). Because RFID is an emerging technology, you should choose the optimum business case and use it as an acid test to validate benefit assumptions. Venturing out with multiple suboptimal choices might dilute the focus and resources of the business.

The first implementation of a business case typically takes the form of a pilot, in which a scoped-out version of the case is attempted. The pilot also represents the first milestone of the business case roadmap. The design and implementation of the pilot should be completed (see Chapter 9, "Designing and Implementing an RFID Solution") within a specific time frame (neither too short nor too long). Pilots tend to be on the shorter side, spanning about eight weeks or so. Depending on scope and complexity, however, the pilot might require a longer time frame for design and implementation. After the pilot has been implemented, its actual benefits must be measured and compared against the business case. With this information in hand, it is possible to fine-tune the business case to align expectations closer with reality. The experience achieved from the pilot implementation serves as a base to design and implement the next iteration—the second milestone of the roadmap. The successful implementation of this iteration, and each subsequent iteration, is then used as a platform on which to build the next tiers of the solution.

The entire business case is thus implemented in a series of iterations rather than a single monolithic chunk. The reasons for this are twofold. First, for each iteration, the implementation and deployment risks, costs, and complexity are kept in check. Second, the impact of the solution on the existing business processes is limited to a particular iteration. Therefore, these factors can be controlled and tackled methodically. In addition, in an extreme situation, if an RFID system fails to deliver on its promises and must be abandoned, a business can do so at minimum cost and with limited impact on the business processes (as compared to a monolithic implementation).

After the business case with the highest priority has been executed, the business case at the next priority level can be implemented in a similar manner. You repeat this process until you have implemented all the required business cases. Figure 7-1 shows the entire flow.

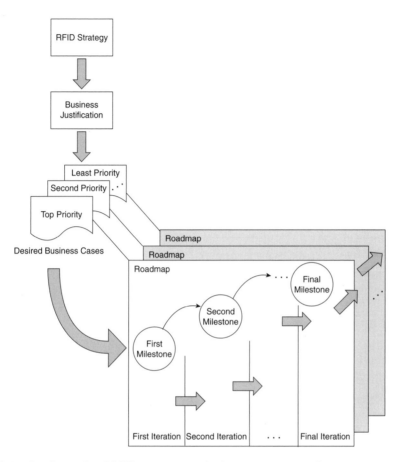

Figure 7-1 A schematic of RFID strategy to deployment strategy flow.

7.4 Conclusion

An RFID strategy, for a large business, provides an enterprise-level plan that aligns with overall business goals and strategies. When formulated at this level, an RFID strategy offers benefits to the widest range of interested parties and facilitates buy-in from senior management. It is strongly recommended that you have an RFID strategy in place before attempting any effort to deploy the technology. You can use an RFID strategy to drive action plans for an enterprise's RFID policy, implementation, and technology rollout.

This chapter has provided some general guidelines and examples to familiarize you with elements of an RFID strategy. A business can select more than one of these guidelines and customize its strategy depending on its core competencies and strategic vision.

Creating Business Justification for RFID

Business justification is strongly recommended before rolling out RFID in an enterprise because it enables you to accomplish the following fundamental goals:

- **Provide objective data about the benefits of RFID.** Objective data enables you to determine whether to use the technology. Indeed, if RFID shows substantial benefits, objective data might accelerate adoption of RFID in the business. Business justification will also build realistic expectations of the technology.

- **Maximize return on investment (ROI).** Through a careful analysis of business cases, you can quantify the benefits and the resources needed to implement an RFID solution. You can select the areas that offer the maximum ROI as the potential candidates for RFID use.

Even if the business has received an RFID mandate from its clients or business partners, it *might* need to justify the use of the technology. After the goals listed here are achieved, it will become apparent how and where RFID needs to be applied. Design and implementation of appropriate RFID solutions can then proceed. In summary, successfully executing these goals will set a solid foundation on which the RFID technology adoption can be built upon.

In general, you cannot determine ROI accurately just by making high-level assumptions regarding benefits. If the shrinkage in a certain type of business accounts on average for 20 percent of sales, for example, a business of this type might be incorrect in assuming RFID will result in savings of 20 percent. First, the actual shrinkage of this particular business might be greater than or lower than this number. Second, introduction of a nontrivial RFID solution in a business operations environment impacts the operations and business flows. Therefore, the final benefit numbers might be off the mark. This important factor shows that an RFID solution cannot be designed in the technical space alone; business impact and analysis need to be performed, too.

An RFID system is a data-collection technology. After the data-collection part is done, you must address the fundamental question of how to use this data. The answer lies in refactoring the business processes so that you can extract maximum advantage from the available data. Therefore, *business process change will offer the ultimate benefit in using RFID technology.* So that you learn how to determine the benefit effectively, this book uses a bottom-up approach. In this method, the business process changes for using RFID are determined first, and then the impact of these changes is analyzed and the benefit determined as a result of this analysis. This approach can lead to accurate determination of ROI and hence realistic expectations regarding the use of the technology.

This chapter provides a practical guide to understanding and analysis of the RFID variables involved in creating the business justifications. This chapter helps you avoid many common and not-so-common mistakes when determining the business justification of your RFID solution.

The business justification method discussed here comprises the following five steps (see Figure 8-1):

1. Forming the business justification team

2. Determining the potential application areas

3. Building business cases

4. Determining priorities

5. Determining roadmaps

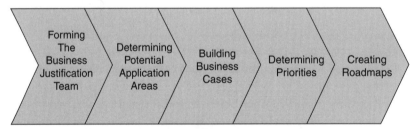

Figure 8-1   A method for determining RFID business justification.

Before delving into a discussion about business justification, you need to understand an important RFID application type called *slap and ship*, which represents probably the most common application type currently implemented by businesses to meet their RFID mandates.

NOTE

This chapter assumes that you are already well versed in general analysis artifacts such as how to create business-flow diagrams, use cases, and so on. Therefore, this chapter does not discuss these topics in detail.

8.1   Slap-and-Ship Type Application

A slap-and-ship type RFID application involves attaching an RFID tag or a smart label to an item and shipping it to the customer who wanted this item tagged. This method represents the simplest way to achieve RFID compliance with minimum investment and implementation complexity. Such an application is either totally separate or minimally integrated with the business processes and back-end enterprise systems. As a result, such an application offers little or no benefit to the business itself that is implementing this application. From another perspective, a slap-and-ship application might be least disruptive to the business operations. It can also provide a company with valuable lessons in RFID should it want to take the application to the next level and integrate it into its business processes. In addition, a slap-and-ship type application provides a starting point for adopting RFID in a business. A business can take its slap-and-ship application and enhance it so that it can also benefit from it (besides just the customer who issued the mandate). A slap-and-ship application is generally not scalable, but is suitable when the volume of tagged items is low. However, it is laborious and susceptible to operator errors. This application type might turn out to be a liability in the long term, especially if competitors can successfully integrate RFID into their business processes to realize its benefits.

Coming back to the discussion of business justification of using RFID, the following sections provide in-depth information on this subject.

8.2   Step 1: Forming the Business Justification Team

An RFID business justification process needs to be performed by a team capable of analyzing the technology benefits for the different facets of the business, which immediately implies a team whose members come from different functional areas of the enterprise. This team should have sufficient RFID knowledge to properly perform the necessary analyses. If the business has already created an RFID strategy and has received buy-in from the senior management, such a cross-functional support might already have been in place (see Chapter 7, "The RFID Strategy"). Even in the absence of an RFID strategy, the formation of such a team is strongly recommended, with its members coming from senior management levels. This team makeup will ensure that all the potential application areas are looked at from proper perspectives. This team can then form subteams, which can include the people both internal and external (for example, consultants and system integrators) with specific knowledge to provide the necessary analysis expertise when and where needed.

8.3   Step 2: Determining Potential Application Areas

You might define several broad categories in which RFID use might improve business efficiency (for example, the manufacturing process, the distribution logistics, administration and security). Only specific areas under these categories should be targeted, however, not the entire category. To improve security, for example, you should pick the areas with urgent security needs (for example, employee pilfering) instead of trying to improve the security of the enterprise as a whole. You can further break down these selected areas to pinpoint the specific spots that promise the most effective impact (for instance, a particular distribution center that accounts for most of the shrinkage). The reason for this is simple. Because RFID is an emerging technology, you must use it selectively and iteratively so that its cost, complexity, and risks (as well as benefits) are kept in check with minimum impact on business operations. Therefore, you should select only a handful of such potential application areas and prioritize them for RFID application. For each such application area, you should develop a business case based on a set of criteria. For example, you might use the following criteria:

- Benefit
- Cost
- Risk
- Complexity
- ROI timeline

The preceding list is not exhaustive; you can supplement it with other business-specific factors. However, these sample factors generally suffice for a business case development. The following sections discuss these criteria in detail (and, thus, the creation of a business case). You can consider these business cases the business justification for using (or not using) RFID in an enterprise.

8.4   Step 3: Building Business Cases

A business case developed for a specific application area is a living and dynamic entity rather than a static, one-time action. Therefore, you need to revisit a business case periodically to account for the impact of the existing factors that might change over time and to incorporate the impact of the new factors (for example, a competitor offering similar but superior capabilities or RFID mandates from new customers and business partners). Also upon review, you can examine a business case to determine how closely it reflects reality (as evident from the associated pilot implementations and proof of concepts). At this point, you can adjudge that the business case correctly captures the impact of its constituting factors.

The following subsections discuss the sample factors mentioned earlier in the section "Step 2: Determining Potential Application Areas."

8.4.1   Benefit

Benefit represents one of the most important factors in building an RFID business case in most any enterprise. You should analyze the benefit factor even if you are planning a slap-and-ship type application. This section discusses a way to quantitatively determine a benefit based on analyzing business flows of the application area under consideration. Other schemes are also certainly possible. This method consist of the following two steps:

1. Creation of business-flow diagrams

2. Determination and analysis of RFID impact points

Apart from quantizing the RFID benefit, this method can also provide some estimates of RFID cost, too. These are discussed next.

8.4.1.1   Creation of Business-Flow Diagrams

The business might not have formal business-flow diagrams for its operations yet. Now is the time to create these artifacts. Several characteristics of the process might become evident just from analyzing these diagrams. Figure 8-2 shows an example business-flow diagram. The nodes in this diagram represent major processing steps in the flow, which in turn can contain several substeps. Each nodes implies a corresponding business use case. Figure 8-3 shows such an example use case. This diagram provides other benefits besides those associated with RFID justification. For example, redundant or inefficient processes or substeps might become evident from a review of the flows and associated substeps. Thus, the business can begin to improve its efficiency even before it has started using RFID for efficiency gains!

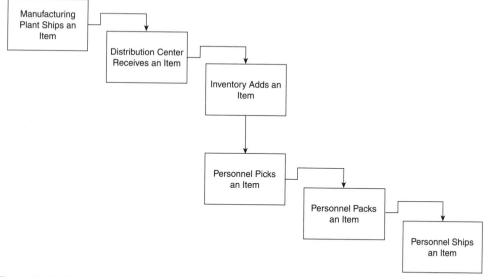

Figure 8-2   An example business-flow diagram.

Use Case Name	Distribution Center Receives an Item
Actor(s)	Receiving personnel (RP)
Triggering Event	Arrival of an item sent from the manufacturing plant (MP) to a distribution center (DC)

1. A shipment containing the item has arrived at a DC Receiving Dock.
2. The pallet contanining the item is unloaded from the delivery truck.

Assumptions

1. A pallet contains 25 items packed invidividually.
2. Estimated Time of Completion (ETC) is the average time it takes to complete a particular step in this use case.

Description

This use case describes the process of a DC receiving an item from the MP.

Total ETC

One pallet with a valid order and no invalid items: (5 + 3*60 + 10*60 + 10*25) seconds = 17 minutes 15 seconds.

Termination Outcome	Condition Affecting Termination Outcome
1. The item is successfully received by the DC RP. 2. The item could not be successfully received by the DC RP.	1. Item absent on the request list (was not ordered).

Major Steps

1. RP scans the bar code order number on the pallet to validate the order. (ETC: 5 seconds)
2. If an invalid order number is found, e.g. the order was not placed by this DC, then the pallet is returned unopened to the MP. (ETC: 2 minutes)
3. Else if a valid order number is found then the RP breaks the pallet. (ETC: 3 minutes)
4. RP scans the individual item bar codes to validate that the pallet contains all the items as per the order. (ETC: 10 minutes)
5. If an item is invalid (i.e. was not part of the order) then RP makes a manual note of this item and sends it back to the MP (ETC: 3 minutes)
6. Else if an item is valid then it is placed on the received area from where it is subsequently moved into the inventory. (ETC: 10 seconds)

Figure 8-3   An example use case.

If these business use cases do not exist, these should be created now (see Figure 8-3). The business-flow diagrams and the associated use cases should not be created in haste. Even if the business has these flows and the use cases currently documented, it is important to validate these with the actual business operations to ensure their accuracy. Also, the flow diagrams and the associated use cases should include all the applicable processing steps and substeps, respectively.

8.4.1.2   Determination and Analysis of RFID Impact Points

The business flows documented in the previous step can now be used to determine the processing steps that RFID will impact. Figure 8-4 shows how the example flow of Figure 8-1 might be affected when RFID is integrated into the business processes.

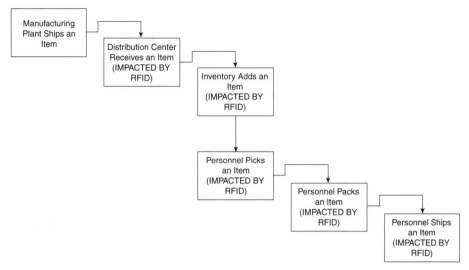

Figure 8-4   An example business-flow diagram impacted by RFID.

Figure 8-5 shows how the example flow of Figure 8-2 might be affected for a slap-and-ship solution.

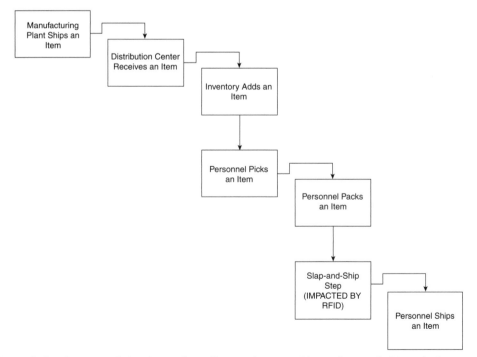

Figure 8-5   An example business-flow diagram impacted by a slap-and-ship solution.

From Figure 8-5, it is evident that the adoption of RFID does not change the existing business processes. One additional step is all that is generally required. Although this might work for the short term, long-term use of this solution is not recommended (as previously pointed out). The following explains one good reason for this. Figure 8-6 shows what the use case associated with the slap-and ship step might look like.

Use Case Name	Slap-and-Ship Step
Actor(s)	Operations personnel (RP)
Triggering Event	Arrival of a packed item ready for shipping

Preconditions
1.   A pallet containing the item has arrived at a DC Shipping Dock.

Assumptions
1.   A pallet contains 25 items packed invidividually.
2.   Estimated Time of Completion (ETC) is the average time it takes to complete a particular step in this use case.

Description
This use case describes the process of performing a slap-and-ship procedure to an item packed on a pallet and ready for shipping.

Total ETC
One pallet with a valid order and no invalid items: $(120 + 25 + 25 + 250 + 500 + 120)$ seconds = 17 minutes 20 seconds.

Termination Outcome	Condition affecting Termination Outcome
1. The item is successfully tagged by the DC RP.	None.
2. The item is successfully tagged by the DC OP.	None.

Major Steps
1.   RP breaks the pallet containing the item (ETC: 2 minutes). This is one time overhead associated with the pallet containing the item.
2.   RP scans the item bar code. (ETC: 1 second)
3.   RFID tag is commissioned for this item. (ETC: 1 second)
4.   RP attaches the this tag to the item. (ETC: 10 seconds)
5.   RP places this item back on the pallet. (ETC: 20 seconds)
6.   If this is the final item on the pallet that was to be tagged then
a.   RP packs the pallet (ETC: 2 minutes). This is one time overhead associated with the pallet containing the item.

Figure 8-6   An example slap-and-ship use case.

If you compare the estimated completion times of the use cases in Figures 8-3 and 8-6, you can see that an additional 17 minutes 20 seconds is incurred when a slap-and-ship type solution is used for a single pallet. If a distribution center ships between 1,000 and 1,500 pallets a day, a cost overhead of 288.9 to 433.3 hours/day is incurred in total. Suppose that the business is spending $10 every hour on labor cost. Then, an amount of $2,888.90 to $4,333.30 will be spent on labor cost alone every day. This translates into a cost overhead of $751,114.00 to $1,126,658.00 per year, assuming a 260-day work year (not including RFID hardware cost, cost of implementation, and so on). This cost could be offset by savings in other areas, such as reduction of shrinkage, increase in shipping accuracy, and so on, not to mention that the company is now complying with the RFID mandate of the largest customer(s). Had the business decided to integrate RFID with its operations, however, it might have saved money together with meeting its customers' RFID mandate (as discussed next).

Adoption of RFID might require new processing steps, and therefore new business use cases (associated with these new steps) might be introduced in the flow. RFID adoption can also lead to introduction of new subprocessing steps for an existing business case. The use cases associated with the impacted and newly introduced processes now need to be analyzed for benefit. How can you discover these impact points? How do you determine whether new processing steps are needed? The answers to these questions lie in a comprehensive understanding of the specific business variables involved and creating a judicious tradeoff among them. The latter depends on the business type and characteristics and is therefore beyond the scope of this book. However, a simple illustration will now be provided as a guide. Figure 8-7 shows how the use case in Figure 8-3 might change. You can estimate the savings, for example, by taking the difference in time taken to complete this and the original use case for one shipment. You can multiply this difference by the number of shipments that a distribution center receives each day. This saved time can then be translated into savings of labor overhead and associated expenses.

Use Case Name	Distribution Center Receives an Item
Actor(s)	Receiving Personnel (RP)
Triggering Event	Arrival of an item sent from the manufacturing plant (MP) at a distribution center (DC)

Preconditions

1. A shipment containing the item has arrived at a DC Receiving Dock.
2. The pallet contanining the item is unloaded from the delivery truck.

Assumptions

3. A pallet contains 25 items packed individually.
4. Estimated Time of Completion (ETC) is the average time it takes to complete a particular step in this use case.

Description

This use case describes the process of a DC receiving an item from the MP.

Total ETC

One pallet with a valid order and no invalid items: (10 + 2 + 20) seconds = 32 seconds.

Termination Outcome	Condition Affecting Termination Outcome
5. The item is successfully received by the DC RP. 6. The item could not be successfully received by the DC RP.	1. Item absent on the request list (was not ordered).

Major Steps

1. RP places the pallet inside the RFID portal. (ETC: 10 seconds)
2. The RFID pallet tag is automatically read by the RFID reader(s).
3. The individual RFID tags are also automatically read by the RFID reader(s).
4. The resulting reads are used to perform an automatic check for order validation. (ETC for steps 2-4: 2 seconds)
5. If an invalid order number is found e.g. the order was not placed by this DC, then the pallet is returned unopened to the MP. (ETC: 2 minutes)
6. Else if a valid order number is found then if an item is invalid (i.e. was not part of the order) then
   a. this information is automatically captured by the system. (ETC: 1 second)
   b. pallet broken by RP. (ETC: 2 minutes, this is a one time overhead)
   c. item is sent it back to the MP. (ETC: 3 minutes)
1. Else if a valid order is found without any invalid item then
   a. then the entire pallet is placed on the received area from where it is subsequently moved into the inventory. (ETC: 20 seconds)

Figure 8-7   An example use case impacted by RFID.

If you compare the estimated completion times of the use cases in Figures 8-3 and 8-7, you can see that a savings of 1,035 − 35 = 16 minutes 40 seconds can be achieved when RFID is used for a single pallet. Therefore, a 96 percent reduction in processing time can be achieved in this case for a single pallet. This is in absence of any exception conditions. If a distribution center handles between 1,000 and 1,500 pallets a day, and 80 percent of the time the conditions are exception free, a savings between 277.78 and 416.68 hours/day can be achieved for this particular use case. Suppose that the business is spending $10 every hour on labor cost. A savings of $2,777.80 to $4,166.80 every day can potentially be achieved on labor cost alone. This translates into an annual cost savings of $722,228 to $1,083,368; assuming a 260-day work year. Individual exception conditions can similarly be analyzed to find out the associated savings.

An in-depth discussion of the business variables constitutes the core material of this section.

To reiterate from the previous discussion, when a nontrivial RFID solution is put into production in an enterprise, it almost always affects its existing business processes. Some subprocesses might get streamlined and thus provide efficiency gains, whereas some other subprocesses might need to include additional processing steps, which might impact their efficiency rates. These extra steps are introduced because of some or all of the following factors:

- Commissioning a tag
- Attaching a tag
- Reading a tag
- Taking corrective measures
- Education and training
- Recycling a tag

You can break down several of the preceding variables further into their constituting subvariables. The following subsections discuss these variables individually.

NOTE
Some of the variables can prove difficult, if not impossible, to analyze using only pen and paper. Designers should be prepared to get their hands dirty by actually using RFID products in trial/pilot setups. This should help them determine the parameters that, in turn, will enable them to make the right business decisions. Chapter 9, "Designing and Implementing an RFID Solution," provides more details on this subject.

8.4.1.2.1   Commissioning a Tag

A tag has to be commissioned (that is, created and uniquely associated with the tagged object record) with some useful data before it can be used in the application. The main questions the designers should ask here are as follows:

- At what points in the existing operations can a tag be created?
- Does a new "tag creation" process step need to be introduced in which the tags will be created or can this be done as a part of an existing process?

If the business is currently producing bar codes at some specific point(s) in its operations where the RFID application is going to be used, these should be the first choice for tag creation points. Although the specific business requirements might not favor using these as tag creation points, nevertheless these should at least be considered before a decision is made. (The goal here is to avoid creating any additional process related to tag creation so as to avoid introducing additional processing steps to the existing operations. Therefore, you might need additional personnel to staff this step and extra time to physically coordinate the created tags to pass on to the next step.) In fact, you should try to combine tag creation with bar code writing as one step. How can you do so? Recall from Chapter 1, "Technology Overview," that an RFID printer can print a smart label, which is a combination of a bar code and RFID tag (together with some free-form descriptive text) in one single physical label. Therefore, if you can supplement existing bar code printers with these printers, you can print both the bar codes and RFID tags in one integrated step.

8.4.1.2.2 Attaching a Tag

Tag attachment is the logical next step after tag creation. However, this step might not *immediately* follow tag creation. Some intermediate subprocess(es) might be executed before a tag can be attached to the appropriate items. Again, designers should ask the following fundamental questions:

- At what points in the existing operations can a tag be attached?
- Does a new "tag attachment" process step need to be introduced in which the tags will be attached or can this be done as a part of an existing process?

If a business is currently attaching bar codes at some specific point(s) in its business operations, these should be the first choice at which to attach the tags. Again, the goal is to avoid creating a separate process to attach a tag, so as to avoid adding a processing step to the existing operations. You might need additional personnel and time to attach the tags to the items, which will most likely increase operational costs that might not be justified in the long run. To avoid getting caught in this scenario, you should combine the tag and the bar code attachment into one step. If the designers have already decided to use smart labels and the business already has a bar code attachment step in its process, you should use this attachment step to attach this smart label, too. Be aware, however, that you might need to attach the smart label with some precision so that the RFID reader(s) can successfully read the embedded RFID tag. For example, if the business has historically just slapped the bar code on items at random spots, you might need to limit their attachment to only specific areas to increase the readability of the attached RFID tag. In the beginning, you might introduce a small delay when attaching the smart label to the item as compared to the time taken originally to attach a bar code label to the same item. However, as workers gets used to properly positioning the smart label on the item, this delay might become negligible. In addition, you can use automatic RFID tag applicators to avoid incurring any delay associated with the tag attachment process.

8.4.1.2.3   Reading a Tag

As important as the tag creation and attachment steps are, the next logical step of reading a tag is equally important. The main questions to ask here are as follows:

- At which points in the operation does the tag data need to be read?
- Do any existing operations need to be altered to do this?

Clearly, the sole reason the business is putting tags on items is because the business wants to read and use the tag data at some point in its operations. A business might consider itself in compliance with "RFID-enabling" mandates by attaching RFID tags to its products and then having some one else (for instance, its customers) read and use the tag data. Even in this scenario, however, a business still needs to read the tag data at some point other than the validation performed by the reader/printer upon tag creation. The additional read(s) will confirm tag validity and whether it contains the relevant data that other parties (its customers) want. Therefore, just slapping an RFID tag on a business product does not mean that the business is RFID-enabled.

At which points (*choke* points) should businesses read the tags? Application requirements determine the "right" points. For instance, is the application used to determine the types of items being loaded into the pallet or the types of items being loaded into the loading truck? In the first case, the tag read might occur at the pallet loading stage, whereas in the latter case it might be done during the truck loading stage (at a dock door, for example).

Next, you must determine whether to alter existing business operation steps at the read point. Suppose, for example, that the tags need to be read when on a moving conveyor on their way to inventory. Do you need to slow down the conveyor so that all tags can be properly read? Will doing so affect business efficiency? It depends. You might find that improved tag read accuracy leads to better inventory management, which in turn might more than offset this efficiency loss. Designers need to use RFID in an experimental setup to determine these factors.

8.4.1.2.4   Taking Corrective Measures

Any RFID solution, even a trivial RFID system, requires implementers to take corrective measures, mainly involving the following two types of actions:

- Corrective actions associated with a bad tag write/attachment/read
- Corrective actions associated with rectifying the improper outcomes of a business operation

The first type of action is required with any RFID system, regardless, for example, of whether Class 0 EPC (tag data is already written by the manufacturer) or Class 0+/1 EPC (tag data is written by the business) tags are used. What happens if the tag is bad or the RFID printer/reader fails to write the tag data completely? The latter might happen, for instance, if a relatively large amount of data needs to be written to the tag (because the chance of this occurring increases with the increase in data size) or the printer may not be configured properly. The only solution to this problem is to write another tag and discard the bad tag, at a cost of additional time to create the new tag and additional resources in the form of manual intervention to discard the tag so that it does not reenter the operation flow. An RFID printer will automatically print a new tag with the same data and mark the old tag as invalid. However, if you cannot attach a valid tag properly to an item, you should strip off this tag from item before attaching another one. Why? Because although a reader cannot read this invalid tag most of the time, some readers might actually be able to read it sometimes. Therefore, if you do not strip off the invalid tag, the item will end up having two unique identifiers. Although the misplaced tag can simply be ignored (for example, by not associating it with the item), the dual (potentially) identifiers might lead to unexpected side effects. For example, suppose a correctly placed tag is put right on the top of this misplaced tag; in this case, the latter might prevent proper reading of the former. Another effect is that the back end might trigger false alarms when a reader reads a misplaced tag. To prevent these occurrences, you should strip off the misplaced tag, and then either attach it properly or use a new tag. Note that in the latter case, you must produce a new tag (adding to the time and resource requirements of the operation). What happens if a reader cannot properly read a tag after you have attached the tag? In this case, you might need to develop auxiliary processes to correct a read problem (additional processing steps). However, the auxiliary process(es) might not mean an overall operative loss. The results gained by properly reading the tags might provide substantial benefit that outweighs the overhead associated with these auxiliary steps.

The determination of improper outcomes of business process(es) is one of the chief benefits of using an RFID system. Suppose, for instance, that tagged cases of a product are being loaded into a truck for delivery to a customer's warehouse. Some cases might be loaded for which the customer has not placed an order. An RFID application can automatically check the shipment validity by reading the tag data from such a case and comparing it with the customer order. Whenever a mismatch is found, the RFID application can trigger a visible or audible alarm to alert the floor personnel. Additional processes must be devised to rectify such mistakes. Although these measures might add time and resource overhead, the aggregate savings achieved by correcting such errors can prove substantial, resulting in significant savings even after factoring in the cost of rectification steps.

8.4.1.2.5   Education and Training

Education and training plays an important part in determining the success or failure of an RFID system; therefore, never overlook this factor when designing a change to any business process. Operations personnel might know different process flows and steps associated with such flows (and might even hold these dear to their heart). For example, if a manual inventory-tracking method (pen and paper) is currently used, you might find it hard to do away with this artifact overnight in favor of an RFID solution that tracks inventory automatically. In addition, you might encounter baseless fear and suspicion that the RFID technology is being used to monitor and control employee activity (which might lead to employee demoralization). To eliminate this fear, you need free and open discussion about the technology. In addition, proper training is essential to reduce hardware damage and system downtime and to ensure the proper use of the system. For example, a forklift driver might unknowingly slam a pallet against the warehouse walls to stack it. As a result, however, the attached pallet tag might be damaged besides (or positioned in such a way that readers cannot read it). You must educate the forklift driver about the impact of his operation style on the RFID system and train the driver in alternative, nondestructive, pallet-stacking procedures. You can also proactively plan training. For example, if you plan to use a new forklift type with the roll out of the RFID system, you can plan operator training in advance. The advance training will enable your forklift operators to perform their duties without disrupting operations or damaging RFID hardware.

8.4.1.2.6   Recycling a Tag

Typical RFID applications do not recycle tags. However, tag recycling can drastically reduce the hardware cost associated with an application and therefore its *total cost of ownership* (TCO). After all, the cost of tags is generally the only recurring hardware cost for any RFID application that does not recycle them. Although tag recycling might seem like a good idea, you must answer the following questions satisfactorily before making a recycling decision:

- What are the business drivers to justify recycling a tag?
- At which points can a tag be retrieved for recycling?
- What steps are required to recycle the tag and subsequently put it back in use?

The business justification for recycling depends on several factors, including the following:

- **Additional personnel required.** You might or might not need additional personnel, depending on the operations' characteristics and recycling strategy.
- **Personnel training.** Although frequently overlooked, some form of training is generally needed for the operations personnel to properly collect and recycle tags. Unless your personnel is properly trained, you risk damage to the tags during the recycling phase.
- **Processing cost.** Typically, some processing cost is associated with a tag (as discussed next).
- **Operations characteristics.** One of these is the physical distribution footprint of the operations. Recycling a tag is more realistic for operations that have small physical distribution footprints than for those that are geographically widely distributed. Tag recycling is more realistic for a business operation housed under a single roof than one that spans several geographic regions.
- **Resources needed to change any business processes.** You must determine whether you need to adjust any existing business processes to accommodate the recycling process. If so, you must calculate the required additional resources (for example, time, money, and personnel) needed to handle these changed processes.

- **Cost of impacting existing business processes.** Do you need to adjust any operations parameters (for example, the rate at which a delivery truck is loaded, or the speed of a moving conveyor)? If so, how will these changes affect the overall efficiency of the operations?

All these listed factors have the potential to add to operating costs. Do the savings from recycling tags outweigh the total cost associated with these factors? If so, the designers should definitely consider the recycling option. If the application is using expensive active (or even passive) tags, designers *should* seriously consider recycling.

Assuming that tag recycling can be justified, at what point in the operation will the tags attached to the items be collected? Generally, you should collect tags at the end of the life cycle of a tagged item. Depending on the definition of a life cycle of an item, this might be when the item is rolling off of an assembly line (in case of an automobile) or when it is ready to be recycled (a chemical drum). This collection will most likely introduce additional processing steps in the business operations. However, the savings achieved via recycling might more than compensate for the overhead associated with these additional steps.

After you have collected the tags, how do you make them usable again? You might think that, at a minimum, you need to erase the data on the tag. However this is *not* true. If the end of the life cycle of the tagged object means that the object is "dead and done" as far as the operations is concerned, you can just reuse the tag identifier on the object for a new, live object that is about to begin its life cycle. Suppose, for example, that a tag with a unique identifier is associated with the *vehicle identification number* (VIN) of an automobile to track it on an assembly line during its build. When the finished vehicle rolls off the assembly line, the vehicle no longer exists (in the context of the assembly line), and therefore this tag can be associated with the VIN of another vehicle that is about to be built. If this tag only contains a unique identifier, it can be reused without any change. However, if this tag contains vehicle-specific data (for example, VIN, custom features such as leather upholstery, color, and so on) relevant to the original vehicle, this data needs to be erased before the tag can be reused. In addition, if the tag requires specialized mounting fixtures, these might have to be fixed so that the tag can be remounted properly on another item. As a result, this step can potentially introduce additional substeps in the processing, too. As mentioned previously, however, savings achieved by recycling the tags might more than compensate for the overhead associated with these additional steps.

8.4.2   Cost

Cost is intimately associated with almost every variable of an RFID system. The following main factors contribute to the total cost incurred in development of an RFID application:

- **Business process changes.** Inherent in any business process change is the potential requirement of additional resources. These resources might take the form of extra personnel or extra time to complete particular tasks(s), resulting in loss of productivity that might directly translate to lost revenue and so on.

- **RFID hardware.** The readers and antennas are a one-time investment associated with an RFID application. However, tag cost can be a recurring overhead if the tags are not recycled. In fact, with regard to an RFID application TCO, tag cost is a major area of concern.

- **Application software and hardware.** Typically, these are one-time investments. However, hardware and software licensing, upgrades, and maintenance costs can be a source of recurring cost.

- **Services.** This has several components. First, the RFID readers and antennas (and controllers) must be connected and set up properly, and the network infrastructure might need to be enhanced to connect the readers. Suitable portals and structures might need to be built to host the readers in the read zones. In addition, the RFID hardware setup might need to be fine-tuned to exploit its maximum potential. Second, the application software might need to be built or customized, and the interfaces to the back-end systems need to be built. Third, testing needs to be done before such an application can be deployed in production. These are only a sample of items that a services component involves. The challenge of accurately estimating the cost of this variable increases with the complexity and scale of the RFID application.

- **Training and education.** This variable generally gets the least attention during the business justification process. However, after the system is up and running in production and the services consultants have moved on to their next projects, the necessity of training to maintain the application might become painfully obvious. You can avoid this "late wake-up call" by building an expert team that you can entrust with such responsibility. In addition, company personnel can "shadow" the services consultants and actively participate with them in the design and implementation of an RFID application.

- **Maintenance.** Typically, the cost of maintaining an RFID application represents the principal share of its TCO. The cost of the tags is a recurring overhead. Even if the tags are recycled, new tags might have to be ordered to replace the ones damaged during recycling. Spare tags, reader antennas, readers, controller hardware, sensors, actuators, and annunciators, among other items, must be ordered to replace defective/damaged hardware. The cost of both vendor and internal technical support, bug fixing, and enhancements is generally an integral part of the maintenance phase. During an RFID system cost analysis, do *not* overlook or underestimate maintenance costs.

Figures 8-8 through 8-11 provide some example cost estimates based on actual RFID projects.

Tags
Item Tagging for a Single Distribution Center

Item Tagging	2004	2005	2006	2007	2008	2009	Comments
Costs of Tags	**$110,400**	**$123,625**	**$125,400**	**$127,050**	**$123,200**	**$113,850**	
Item - Total Tags	230,000	287,500	330,300	385,000	440,000	495,000	
Item - Total Tag Cost	111,400	123,625	125,400	127,050	123,200	113,850	
Initial Tag Requirements - Item							
Number of items received per year	200,000	250,000	300,000	350,000	400,000	450,000	The number of items received increases as operation efficiency is expected to increase
Recurring Tag Requiremnents - Item	**30,000**	**37,500**	**30,000**	**35,000**	**40,000**	**45,000**	
% of items lost/damaged each year	0%	0%	0%	0%	0%	0%	Assuming that there are no damaged or lost items.
% Expected defective tags	15	15	10	10	10	10	This is an estimate, the actual percentage may be higher
Cost per passive UHF tag	$0.40	$0.35	$0.30	$0.25	$0.20	$0.15	Only passive tags required, cost assumes increased adoption
Tagging Costs - (5% tag cost)	$0.08	$0.08	$0.08	$0.08	$0.08	$0.08	Labor cost: $10USD / hour * 30 seconds per tag
Cost per Item Tag	**$0.48**	**$0.43**	**$0.38**	**$0.33**	**$0.28**	**$0.23**	Tag + Tagging Cost
Total Cost (Recurring):	**$14,400**	**$16,125**	**$11,400**	**$11,550**	**$11,200**	**$10,350**	
Total Cost (One-time):	**$96,000**	**$107,500**	**$114,000**	**$115,500**	**$112,000**	**$103,500**	

Key Assumptions
1. One tag per item
2. UHF Passive / WORM tags are used
3. Cost for Passive tags decreasing over time as estimated by the tag vendor

Case Tagging for a Single Distribution Center

Item Tagging	2004	2005	2006	2007	2008	2009	Comments
Costs of Tags	**$11,040**	**$12,363**	**$12,540**	**$12,705**	**$12,320**	**$11,385**	
Case - Total Tags	23,000	28,750	33,000	38,500	44,000	49,500	
Case - Total Tag Cost	11,040	12,363	12,540	12,705	12,320	11,385	
Initial Tag Requirements - Case							
Number of cases received per year	20,000	25,000	30,000	35,000	40,000	45,000	The number of cases received increases as operation efficiency is expected to increase
Recurring Tag Requirements - Case	**3,000**	**3,750**	**3,000**	**3,500**	**4,000**	**4,500**	
% of cases lost/damaged each year	0%	0%	0%	0%	0%	0%	Assuming that there are no damaged or lost cases.
% Expected defective tags	15	15	10	10	10	10	This is an estimate, the actual percentage may be higher
Cost per passive UHF tag	$0.40	$0.35	$0.30	$0.25	$0.20	$0.15	Only passive tags required, cost assumes increased adoption
Tagging Costs - (5% tag cost)	$0.08	$0.08	$0.08	$0.08	$0.08	$0.08	Labor cost: $10USD / hour * 30 seconds per tag
Cost per Case Tag	**$0.48**	**$0.43**	**$0.38**	**$0.33**	**$0.28**	**$0.23**	Tag + Tagging Cost
Total Cost (Recurring):	**$1,440**	**$1,613**	**$1,140**	**$1,155**	**$1,120**	**$1,035**	
Total Cost (One-time):	**$9,600**	**$10,750**	**$11,400**	**$11,550**	**$11,200**	**$10,350**	

Key Assumptions
1. One tag per case
2. 10 items per case

Pallet Tagging for a Single Distribution Center

Item Tagging	2004	2005	2006	2007	2008	2009	Comments
Costs of Tags	**$552**	**$618**	**$627**	**$635**	**$6,160**	**$569**	
Pallet - Total Tags	1,150	1,438	1,650	1,925	22,000	2,475	
Pallet - Total Tag Cost	552	618	627	635	6,160	569	
Initial Tag Requirements - Pallet							
Number of pallets received per year	1,000	1,250	1,500	1,750	20,000	2,250	The number of pallets received increases as operation efficiency is expected to increase
Recurring Tag Requiremnents - Pallet	**150**	**188**	**150**	**175**	**2,000**	**225**	
% of pallets lost/damaged each year	0%	0%	0%	0%	0%	0%	Assuming that there are no damaged or lost pallets.
% Expected defective tags	15	15	10	10	10	10	This is an estimate, the actual percentage may be higher
Cost per passive UHF tag	$0.40	$0.35	$0.30	$0.25	$0.20	$0.15	Only passive tags required, cost assumes increased adoption
Tagging Costs - (5% tag cost)	$0.08	$0.08	$0.08	$0.08	$0.08	$0.08	Labor cost: $10USD / hour * 30 seconds per tag
Cost per Pallet Tag	**$0.48**	**$0.43**	**$0.38**	**$0.33**	**$0.28**	**$0.23**	Tag + Tagging Cost
Total Cost (Recurring):	**$72**	**$81**	**$57**	**$58**	**$560**	**$52**	
Total Cost (One-time):	**$480**	**$538**	**$570**	**$578**	**$5,600**	**$518**	

Key Assumptions
1. One tag per pallet
2. 20 cases per pallet

Figure 8-8   An example estimate of tag costs associated with a project.

Readers
For A Single Distribution Center

	2004	2005	2006	2007	2008	2009	Comments
Total Facilities to Date	3	6	10	15	21	32	
Receiving Docks	2	2	3	4	5	6	
Storage Rooms	1	1	1	1	1	3	
Readers							
Handheld Readers for Receiving Docks	2	2	3	4	5	8	One Handheld Reader for each Receiving Dock
Fixed Readers for Receiving Docks	2	2	3	4	5	8	One Fixed Reader for each Receiving Dock
Antennas per reader	2	2	2	2	2	2	
Fixed Readers for Storage Rooms	1	1	1	1	1	3	One Fixed Reader for each Storage Room
Antennas per reader	2	2	2	2	2	2	
Number of Damaged Handheld Readers replaced	1	1	1	2	2	2	Estimated - actual number could be different
Number of Damaged Fixed Readers replaced	1	1	1	2	2	2	Estimated - actual number could be different
Number of Damaged Antennas replaced	2	2	2	4	4	6	Estimated - actual number could be different
Cost per Handheld Barcode/RFID Reader	$4,500	$4,000	$3,500	$2,500	$2,000	$1,500	Handhelds do not require separate antenna
Cost per Fixed Reader	$1,500	$1,200	$1,000	$800	$700	$500	Estimated based on vendor input
Cost per Antenna	$250	$200	$150	$125	$110	$100	Estimated based on vendor input
Subtotal for Readers & Antenna	**$13,500**	**$11,600**	**$14,700**	**$14,450**	**$14,820**	**$18,200**	
Wiring Costs - Facility	**$2,700**	**$2,700**	**$3,900**	**$5,100**	**$6,300**	**$10,500**	
Wiring Costs per Receiving Dock	$1,200	$1,200	$1,200	$1,200	$1,200	$1,200	
Wiring Costs per Receiving Dock	$300	$300	$300	$300	$300	$300	
Total Cost:	**$16,200**	**$14,300**	**$18,600**	**$19,550**	**$21,120**	**$28,700**	

Assumptions:

1. A tag will be already attached to the item/case/pallet when it is received at the Distribution Center

Figure 8-9 An example estimate of reader costs associated with a project.

Infrastructure Costs
For a Single Distribution Center

	2004	2005	2006	2007	2008	2009	Comments
Hardware Costs	$3,500	$3,500	$6,000	$7,500	$8,000	$10,000	
Total PCs	1	1	2	3	4	5	Redundancy needs to be added starting 2006.
Cost per PC	$3,500	$3,500	$3,000	$2,500	$2,000	$2,000	3.0 GHz Pentium processor, 2 GB RAM, 120GB HD, Windows 2003 Server OS
Software Licenses	$12,000	$12,000	$24,000	$36,000	$48,000	$60,000	
Middleware License/Host	$7,000	$7,000	$7,000	$7,000	$7,000	$7,000	
Database License/Host	$5,000	$5,000	$5,000	$5,000	$5,000	$5,000	
Total	$15,500	$15,500	$30,000	$43,500	$56,000	$70,000	

Figure 8-10 An example estimate of software costs associated with a project.

Software Integration & Application Implementation Costs

For a Single Distribution Center

	2004	2005	2006	2007	2008	2009	Comments
Analysis and Design							
- Pilot and Targeted Launch*	$ 15,000	$ -	$ -	$ -	$ -	$ -	
Go-Live Support for Targeted Launch	$ 15,000						
Construction / Pilot	**$ 23,000**	**$ 10,000**	**$ 10,000**	**$ 10,000**	**$ 10,000**	**$ 10,000**	
Initial Development	$ 18,000	$ -	$ -	$ -	$ -	$ -	
Modification Cost	$ -	$5,000.00	$5,000.00	$5,000.00	$5,000.00	$5,000.00	
Training per Facility (RFID)	$ 5,000	$ 5,000	$ 5,000	$ 5,000	$ 5,000	$ 5,000	Standard cost per facility for training
Middleware and Application Database	**$ 100,000**	**$ 42,000**	**$ 42,000**	**$ 42,000**	**$ 42,000**	**$ 42,000**	
Initial Middleware Customization	$ 25,000						
Modification Costs	$ 10,000	$ 10,000	$ 10,000	$ 10,000	$ 10,000		
Initial Middleware Reports	$ 10,000						
Modification Costs $5,000		$ 5,000	$ 5,000	$ 5,000	$ 5,000	$ 5,000	
							High Complexity - 4 Databases: 1) Order Database 2) Shipment Database 3) Item Database 4) RFID Application Database*
Database Development	$ 30,000						
Modification Costs		$ 12,000	$ 12,000	$ 12,000	$ 12,000	$ 12,000	
Standalone Applications Development	$ 30,000						Low Complexity
Modification Costs		$ 10,000	$ 10,000	$ 10,000	$ 10,000	$ 10,000	
Training per Facility	$ 5,000	$ 5,000	$ 5,000	$ 5,000	$ 5,000	$ 5,000	
Interface Costs	**$ 15,000**	**$ 9,000**	**$ 9,000**	**$ 9,000**	**$ 9,000**	**$ 9,000**	
Number of Interfaces	3	3	3	3	3	3	
Cost per interface	$ 5,000	$ 0	$ 0	$ 0	$ 0	$ 0	Low complexity
Modification Cost per interface	$ -	$ 3,000	$ 3,000	$ 3,000	$ 3,000	$ 3,000	Low complexity
Total Cost:	**$168,000**	**$61,000**	**$61,000**	**$61,000**	**$61,000**	**$61,000**	

Figure 8-11   An example estimate of services costs associated with a project.

Cost is probably the most important variable of all, but it can prove difficult to estimate this variable accurately because of the many variables involved. In the heat of "RFID-enabling," companies might underestimate the cost and overestimate the benefits associated with such an effort. One way to avoid this situation is to start small and achieve "RFID-enablement" by approaching the final solution in a series of controlled iterations and use the experience gained at successive iterations. By doing so, you can constrain the cost at each step, which will result in better cost accuracy and estimation compared to if you just build the entire application in a single monolithic step.

8.4.3   Risk

Risk is inherent in any real-world project; only the degree of risk varies. When assessing the risk of using RFID in an application area, you must consider how the introduction of the technology will impact the business. The amount of expected downtime in operations to rectify hardware issues, side effects of a bad tag read operation, the consequences of not being able to read a tagged item in situations such as when it is packed to deep inside a pallet, your customers' perception of the technology—all these can have a substantial bearing on the company bottom line, customer loyalty, and reputation. In certain situations, although the bottom-line risk might be low, the risk of tarnishing a brand image might prove to be a risk that a company can ill afford. How can you determine risk objectively? An empirical way to do this is to evaluate the application risk point scale of 1 to 5 (where 1 represents *very low*, and 5 represents *very high*). Input from the members of the cross-functional team, together with the guidance from the experts, can help you to accurately determine the risk involved in an application area.

8.4.4 Complexity

The complexity involved in an RFID application can be hard to estimate theoretically. You can gain a general idea of the complexity from the requirements (for example, from the business flows and use cases) and scope of the application. To validate the complexity assumption, however, hands-on experimentation will most likely be necessary in the form of proof of concepts. At this point, you might need the help of vendors, experienced integrators, and consultants. The implementers and the experts should, in the end, agree on the evaluation. You can use the empirical scale used for measuring the risk of a business case when measuring complexity, too.

8.4.5 ROI Timeline

Although an RFID system might offer solid benefit within a tolerable cost, risk, and complexity range, the ROI timeline might prove substantial. Various factors impact ROI, including obtaining government/regulatory body/union approval, receiving consensus from the important stakeholders, determining and following bureaucratic procedures for necessary clearance, and so on. Whereas an ROI timeline for an application area might be as little as a couple of months, it might also be more than one year for certain applications.

8.5 Step 4: Determining Priorities

The factors involved in building a business case can now be combined to determine its priority among the set of all business cases created for RFID adoption. To get a relative sense of how the different business cases of the selected application areas stand, you can pit these against themselves in different ways. Several multidimensional visualization tools you can use for this purpose are currently on the market, and some are in the public domain.

A simple way to visualize the variables without using any such tools is to create a super "opposing" factor by combining the cost, risk, and complexity factors. One way to calculate such a variable is to take the simple average of these factors. For example, if benefit, cost, risk, and complexity factors are measured using a point scale of 1 to 5, as described previously, the "opposing" factor for a business case whose cost = 2.5, risk = 2, and complexity = 2 is (2.5 + 2 + 2)/3, or 2.17 (approximately). You can also calculate a suitable weighted average. Note that if the factors have extreme values (for example, cost = 4, risk = 1, complexity = 1), such a variable can skew the measures somewhat. The benefit, "opposing" factor, and ROI timeline can then be plotted in a three-dimensional coordinate system to visualize the different parameters. This generally clusters the business cases into distinct groups or vertical *cones*. Each such cone represents a certain priority and can also be called *priority cones*. The business cases inside a priority cone can again be ranked against each other to assert their relative priorities. It is suggested that business cases with relatively good benefit and shorter ROI timeline with acceptable cost, low complexity, and risk be given preference for implementation and thus the highest priority unless the business has good reasons otherwise. The priority cones for these business cases tend to be closest to the origin of the coordinates.

Figure 8-12 shows an example set of business cases with benefit, cost, risk, complexity, and "opposing" factor (using a simple average of cost, risk, and complexity) all plotted on a scale of 1 to 5, and ROI timelines in months. Figure 8-13 shows the resulting priority cones for these use cases. The business cases inside the priority cone P should be assigned the highest priority and implemented first.

Case Number	Business Case Name	Benefit	Risk (R)	Cost (M)	Complexity (C)	Opposing Factor = (R+M+C) / 3	ROI Timeline
A	Reduce Labor Cost	4	1	2	2	1.67	3
B	Reduce Shrinkage	4	2	2	2	2	3
C	Automate Receiving	3	2	2	3	2.3	6
D	Increase Security	3	2	3	3	2.67	6
E	Improve Inventory Management	4	3	4	4	3.67	9
F	Reduce Order Turnaround Time	3	4	4	5	4.3	9
G	Increase Customer Satisfaction	3	4	3	4	3.67	12

Figure 8-12   An example comparison of the factors of a group of business cases.

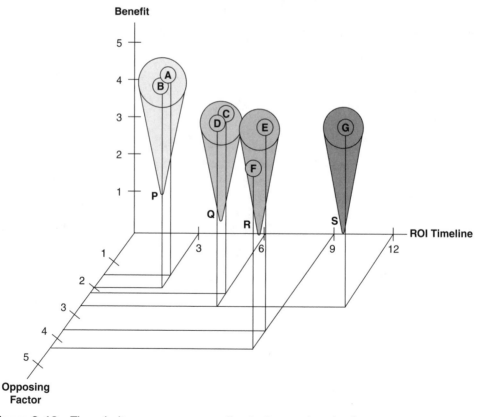

Figure 8-13   The priority cones corresponding to the previous business cases.

8.6   Step 5: Creating Roadmaps

Now that a set of business cases have been selected based on their priorities, the last task is to create a roadmap for each such business case. A roadmap breaks up the complete solution of a business case into a set of milestones. Each milestone corresponds to an iteration of the final solution. It contains a strategically planned subset of the entire scope of the business case that can be achieved within the time frame of the milestone. The design and implementation effort corresponding to a particular milestone typically begins when the previous milestone has been completed successfully.

Consider a concrete example: Suppose that a business case named Track Inventory Using RFID calls for an RFID solution to ship orders from four *distribution centers* (DCs), all using the same operation flow. Also assume that each DC has 10 dock doors. Figure 8-14 shows what an example roadmap for this business case might look like.

Business Case Name		Track Inventory Using RFID
Roadmap		
Milestone One	**Description**	Development of a controlled pilot for 2 dock doors of DC 1
	Duration	8 weeks
Milestone Two	**Description**	Expansion of the pilot to include 4 more dock doors of DC 1
	Duration	8 weeks
Milestone Three	**Description**	Expansion of the pilot to include all 10 dock doors of DC 1
	Duration	8 weeks
Milestone Four	**Description**	Replication of the pilot in DC 2
	Duration	12 weeks
Milestone Five	**Description**	Replication of the pilot in DC 3 and DC4
	Duration	20 weeks

Figure 8-14   An example roadmap for a sample business case.

Note how methodically the scope (of the complete RFID solution as targeted by the business case) is broken down so that each successive iteration builds upon the successful completion of the preceding one. Also note that the first three iterations are small in scope, which is deliberate so that the principal impacting factors of the solution can be identified and solved while keeping their effect constrained within a single DC. After this has been completed successfully, the fourth iteration attempts to replicate the solution in the second DC. A new set of variables might show up associated with the duplication effort that was not apparent in the first three iterations. After this has been tackled successfully, the last iteration duplicates the solution in the remaining two DCs.

8.7   Conclusion

Most likely, RFID will *not* offer every business compelling reasons to adopt it in the near term. However, the decision to use or not to use RFID should not be determined subjectively based on the hype or sweeping negative opinions about the technology. This chapter examined a method to objectively analyze business parameters to determine why and where RFID should be used. If the business is already facing an RFID mandate from its major customers, the decision to use RFID most probably has already been made. The business could use a slap-and-ship approach to meet this mandate while minimizing the time, effort, and resources required. However, the optimum benefit from RFID will be realized only when the business itself assimilates and uses the technology. To this end, the business can at least initiate the business analysis needed to justify using RFID for its own use. It might be pleasantly surprised at the result.

CHAPTER 9

Designing and Implementing an RFID Solution

Just how challenging can it be to design and implement a nontrivial RFID solution? Someone who has not implemented such an RFID system might think, "Not much at all! After all, what else do you need besides a few readers, antennas, cables, and some tags to build any RFID system?" The short answer is this: plenty. This chapter provides the long answer. Suffice to say that designing and implementing a real-world, nontrivial RFID solution is not easy. Therefore, if you are expecting to use a plug-and-play RFID solution for your business needs, be forewarned: The unique needs of each business and the involvement of several variables influence the appropriateness of an RFID solution. No single one-size-fits-all RFID solution exists. Depending on your business needs, you can find several solution components commercially available today from hardware and software vendors as well as integrators. The task is to know which of these components will provide the optimum solution and how you need to put these elements together to achieve the desired solution.

This chapter provides designers and implementers with an in-depth examination of the variables involved in an RFID solution. The chapter also covers the complexity and potential pitfalls involved in designing and implementing a nontrivial RFID solution. In addition, you will find several practical tips and suggestions based on real-world design and implementation experience. From this chapter, you can take away a rich set of tools and guidelines to use to craft your RFID system.

> **NOTE**
>
> This chapter assumes that the designers and implementers are already well versed in general application design and implementation, such as crafting system architecture and so on. It also assumes that the appropriate business use cases, application requirements, and nonfunctional requirements activities have been completed at this point. Therefore, this chapter does not discuss these topics in detail.
>
> Some of the topics covered in this chapter, such as vendor selection, might seem beyond the context, in a traditional sense, of system design and implementation. However, these seemingly out-of-scope variables have important bearing on design and implementation assumptions, and are therefore considered in this chapter.
>
> Some of the variables can prove difficult if not impossible to analyze using only pen and paper. Designers and implementers should be prepared to get their hands dirty by actually using RFID products in trial/pilot setups. This hands-on experience should help them determine the parameters that, in turn, will enable them to make the right technical decisions.

9.1   System Architecture

As a part of the design activity, you must determine the logical architecture of the system. You can use the logical architecture of an RFID system shown in Figure 9-1 (derived from Figure 1-4 in Chapter 1, "Technology Overview") as a starting point.

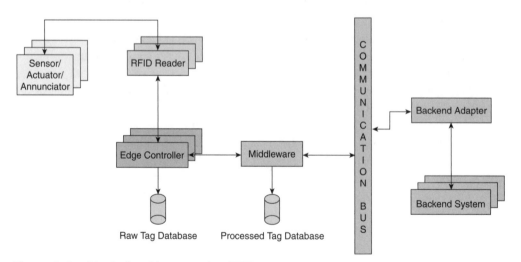

Figure 9-1   A logical architecture of an RFID system.

You can customize this generalized architecture based on application requirements. For example, Figure 9-2 provides one possible customization for a slap-and-ship type application.

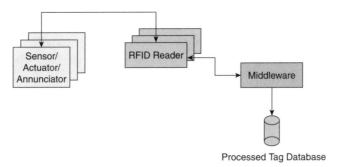

Processed Tag Database

Figure 9-2   A logical architecture of a slap-and-ship type application.

Note a few things in Figure 9-2: The Edge Controller is assumed to be embedded into the network reader itself. The Raw Tag Database is assumed to be present in the reader itself in the form of some kind of read buffer feature. This database can also be hosted by the Middleware component itself. The Middleware can create the raw tag records in this database as soon as it receives a list of read tags from the associated reader (that is, before processing this data). In addition, the processed tag data is not shared with the corporate back-end applications. However, indirect sharing using the Processed Tag Database directly is a possibility. Such a scheme would probably use a batch process to extract the data from this database periodically. The extracted data can then be used in various manners to feed into the corporate back-end systems. This is not a recommended long-term solution, but can be used as a short-term one. The logical architecture of the system now can be used as a blueprint to determine which components are needed and how they need to be connected to implement the solution. For example, the physical architecture of the system can now be determined. Figure 9-3 shows a physical architecture that has been derived from the logical architecture shown in Figure 9-2. (Exact product details like vendor name, part number, and so on are not shown to maintain vendor agnosticism. But these details should be part of a physical architecture.)

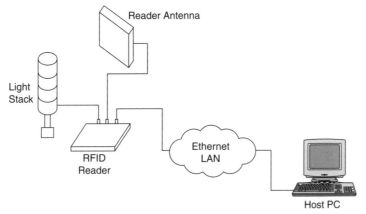

Figure 9-3   A physical architecture of a slap-and-ship type application.

In general, you can determine all the components of a physical architecture only after thoroughly analyzing the RFID technical variables. The next section covers this topic.

9.2   Technical Variables

You must consider several technical variables when designing and implementing an RFID solution, including the following:

- Frequency of operation
- Tags
- Readers
- Antennas
- Items to be tagged
- Operating conditions
- Vendors
- Standards
- Application software and hardware
- Integration with existing systems
- Maintenance

Several of these variables further consist of subvariables. An optimal RFID solution has to determine a good tradeoff among these variable to arrive at a satisfactory working solution that is within the budget and schedule—not an easy task by any means. This task becomes increasingly challenging with the increase in scale and complexity of the solution.

This chapter provides sample templates that you can use to methodically analyze these variables and record your findings. You can customize these templates to suit your particular requirements. You might simplify a template provided here by deleting variables not pertinent to your project. On the other hand, you can enhance a template by adding other custom variables so that the template covers your particular project variables adequately. A template can refer to other templates. For example, the Item ID field of the Tag template refers to the Item ID field of the Item template. A template provided here captures the RFID-specific variables and their subvariables. It does not generally capture other common project-related variables such as personnel, although that is accounted for in the Maintenance template. In addition, you can further normalize a template in the sense that you can extract a set of fields from it and make those into an independent template. For example, Readability fields in the Tag template can be taken out and combined with the Vendor and Test Case templates to create a Readability template. This chapter leaves these and other possibilities to you to explore so that the discussion can focus on RFID-specific variables.

The following subsections discuss each of the previously listed variables.

9.2.1   Frequency of Operation

Figure 9-4 provides a sample template for this variable.

Variable Name	Frequency		
Selection Summary			
Value Options	☐ LF: ☐ 125KHz ☐ 134KHz ☐ Other:_____ ☐ HF: ☐ 13.56MHz ☐ Other:_____ ☐ UHF : ☐ 303.8 MHz ☐ 433 MHz ☐ 868 MHz ☐ 915 MHz ☐ Other:_____ ☐ Microwave: ☐ 2.45 GHz ☐ 5.8 GHz ☐ Other:_____		
Constituting Variables			
Maximum Reading Distance	Priority		
Application Type	Priority		
Operating Conditions	Priority		
Notes			

Figure 9-4   Frequency template.

Frequency of operation is by far the *most important technical variable* in designing an RFID system. Designers should not proceed any further with technical design work until they have determined the exact frequency of operation of a system. Why does frequency play such an important role? The answer is simple: The capabilities of RFID systems vary widely depending on the frequency type. After you know the frequency type, you can methodically tackle all the remaining variables.

How can you determine the frequency of a system? The following three primary factors typically enable you to do so:

- **Maximum reading distance.** This is the maximum distance between a reader and a tag at which the tag data can be read accurately (by the application and depending on the tag type, whether active or passive).

- **Application type.** Generally, well-defined classes of applications are associated with a specific frequency range.

- **Operating conditions.** Some characteristics of the operating environment can be of great help in pinpointing the frequency.

Before delving into these factors in detail, you need to understand an important nontechnical issue regarding frequency selection: the legal limitations on the frequency that the designers have selected. If a 915 MHz frequency has been selected for an RFID application that needs to be deployed in Europe, for example, you have a problem because this frequency is not a legal European frequency for RFID. You can, however, use the 868 MHz frequency instead. In contrast, if you select a 868 MHz frequency for an RFID application to be deployed in the United States, again you have a problem because this is *not* considered legal in America. So what can you do if a selected frequency is not legal in the region of deployment? You have the following options:

- You can change the frequency to the closest equivalent permissible in the deployment location. For example, if 915 MHz is chosen originally for Europe, you can change this to 868 MHz, which is still a UHF frequency and as such will not have a substantial impact on the design.

- Select only those vendor(s) who can provide RFID readers, antennas, and tags that operate in the selected frequency.

- Apply for a special license from the government or appropriate governing body.

The first option involves changing the illegal frequency to its closest equivalent that is permissible in the deployment region and at the same time satisfactorily meets all the application requirements. This is probably the easiest to do, assuming, of course, such a permissible frequency is available.

Vendors can also provide a working solution to this problem. Several current vendors now have readers that can operate in multiple frequencies, thus bypassing the legal restrictions on frequencies. This option also proves useful if a single RFID application needs to be deployed in different geographical regions and the frequency band is illegal in at least one of the regions. If you need to deploy a UHF frequency application in the United States and Europe, for example, you should select those vendors who can supply RFID hardware that can operate in both 915 MHz and 868 MHz. If you do so, the same application will work in both the places with little or no change, and thus alleviate the implementation and maintenance nightmares associated with creating two different systems that are functionally identical.

Some countries grant special licenses for temporary use of a frequency band, even though this frequency might be illegal in that country. However, this approval process might prove time-consuming. If you do pursue this option, allow yourself plenty of lead time. This option proves useful in certain situations, such as if you are planning a pilot study for a temporary period of time to evaluate the technology or validate some assumptions made about certain aspects of the technology. Long-term special license for an illegal frequency might not be easy to obtain unless the RFID application is one of a kind that meets a dire need in some important sectors, such as national defense.

The following subsection explores the factors that can determine the right frequency for an RFID application.

9.2.1.1 Maximum Reading Distance

Maximum reading distance is one of the most important factors in determining the frequency of a system. The reading distance generally narrows down the range of possible frequencies to one or a couple, out of which it is relatively easy to determine the best one. For example, for solutions that need close proximity of the tag (from a few inches to less than 3 feet) to the reader, you should use either LF (typically 125 KHz or 134 KHz) or HF (typically 13.56 MHz). Chapter 1 discusses these frequency types in detail.

9.2.1.2 Application Type

In general, each frequency range is associated with classic applications. For example, for proximity card type applications, LF (typically 125 KHz) is typically used. So if a proximity card type application is being designed, the designer can select the 125 KHz frequency by just matching the application type with its appropriate frequency (see Chapter 4, "Application Areas").

9.2.1.3 Operating Conditions

Some typical operating conditions can provide a strong hint to the frequency. For example, if the application needs to be less affected by things such as metal, mud, or snow, an LF (typically 125 KHz) or an HF (typically 13.56 MHz) is probably the ideal choice. Also, in hospitals, HF (13.56 MHz) is the typical frequency of choice because it poses minimum interference issues with the existing equipment.

The reading distance, generally, is the ultimate deciding factor in frequency selection. Although the application type and operating conditions are helpful, it is the reading distance that finally determines the correct frequency type. What can you do if a frequency selected based on the reading distance differs from the one selected based on other factor(s)? In this case, the factors should be prioritized in terms of importance. The frequency associated with the most important factor should then be selected. For example, one large tire manufacturing company deviated from the Automotive Industry Action Group's B-11 standard for North America to build its application to monitor tire pressure and temperature. The B-11 standard specifies a reading distance of 24 inches and a frequency of 915 MHz. However, the operating conditions distorted the results, and the company decided to use the LF (125 KHz) instead so that its application would be less impacted.

9.2.2 Tags

Figure 9-5 provides a sample template for this variable.

Variable Name	Tag		
Selection Summary			
Constituting Variables			
Type	☐ RO ☐ WORM ☐ RW		
	☐ Passive ☐ Active ☐ Semi-active		
	Special Characteristics		
	Other		
Readability	Item ID		
	Read Robustness		
	Read Distance		
	Tag Density		
	Tag Motion		
	Tag Orientation		
	Operating Environment		
	Item ID		
	Read Robustness		
	Read Distance		
	Tag Density		
	Tag Orientation		
	Operating Environment		
Data Capacity	Tag Motion		
	Object Data		
	Data Lock?	☐ Yes ☐ No	
Physical Properties	Dimension		
	Ruggedness		
Attachment	Item Type/Name		
	Where?		
	How?		
	Item Type/Name		
	Where?		
	How?		
Tag Recycled?			
Volume			
Notes			

Figure 9-5   Tag template.

This seemingly simple component involves the following subvariables, virtually all of which you need to considered when designing an RFID system:

- Type
- Readability
- Data capacity
- Physical properties
- Attachment
- Volume

Optimizing each of the variables might turn out to be an extremely difficult task, if not an impossible one. You should first attempt to achieve the optimum balance among the variables. If you cannot do so, make an acceptable tradeoff. This is one of the decisions that you should not take lightly or make in haste, because this is the single most important decision that will influence the recurring application cost and the data-collection capabilities of the application.

The following subsections examine each these variables.

9.2.2.1 Type

The type variable refers to the tag type that will be used in the application. The following are some of the major questions that you need to answer in this context:

- RO, WORM, or RW?
- Passive, active, or a semi-active?
- Special characteristics?
- Other?

As you can understand from this list, even a simple-looking decision about tag type involves several variables that you need to carefully consider.

9.2.2.1.1 RO, WORM, or RW?

Will the tag already have a unique identifier written onto it by the manufacturer, or is the application going to create the tag when it needs one? Can this tag data be rewritten after it is created? If so, how is data security addressed to make sure that the tag data is not changed unscrupulously by unauthenticated entities? The answers to these questions depend on the type of data that needs to be stored on the tags. If the tag just needs to store a unique static identifier, a RO or a WORM tag is exactly what is needed. If the tag data can change, which means that it contains dynamic data specific to the tagged object (together with a unique static identifier), RW tags are needed. If the application is going to be used inside the four walls of an enterprise with acceptable security in place, then *perhaps* security is not a great concern. However, you should evaluate this based on the company's auditing capability, shrinkage history, and so on.

9.2.2.1.2   Passive, Active, or Semi-Active?

Generally, unless the application needs custom tag features—such as sensing the temperature, humidity, and so on—that differ from the basic tag properties of storing and transmitting its data, passive tags might do just fine. However, the RFID system will definitely need active tags if such custom features are required that need on-board processing. In addition, if the application involves high-speed movement of tagged objects, use of semi-active tags is a possibility.

9.2.2.1.3   Special Characteristics?

Special characteristics (for example, tags customized for tagging metallic objects or items containing RF-absorbent liquids) might be required depending on the application requirements. This is an important concern. System performance could be seriously degraded if an incorrect tag type is chosen that leads to poor readability when attached to the desired object. Note that a suitable tag for tagging the item in question might not exist! Developing a custom tag for a particular item is generally expensive and typically costs $100,000 or more.

9.2.2.1.4   Other?

The tag might have additional properties that are not captured by the previously mentioned subvariables. If EPC tags are used, for example, this subvariable can mean EPC Class0/0+/1 or EPC Gen 2. In addition, if a particular tag type from a vendor is used, it can be associated with this variable together with any vendor-specific product code number.

9.2.2.2   Readability

Readability is an extremely important variable associated with a tag, because you want a guarantee that a tag will be successfully read by a reader in the operating environment. Therefore, this element alone can determine the success or failure of an RFID system. To ascertain the readability of a tag, consider the following:

- **Read robustness.** Tag read robustness refers to how many times this tag can be read when attempted by a reader. The higher this number, the higher the robustness. Thus, a good read robustness means that there is a high possibility that a tag will be read at least once by a reader. As a result, the resulting RFID system will have a high degree of successful data capture from the tags.

- **Reading distance of a tag from a reader.** The tendency to maximize the reading distance should be avoided if this has a negative impact on the read robustness. An RFID system should be designed so that it optimizes the read robustness with the reading distance requirement.

- **Tag antenna design.** This factor has tremendous influence on tag readability. A tag antenna can be designed in literally infinite ways. A tag antenna whose design has been optimized to provide certain characteristics under a specific set of operating conditions might perform poorly under a different set of conditions. The hardware vendor is responsible for tag antenna design. Therefore, the business does not have much control over this factor.

- **Tag density.** This factor denotes how many tags are present in the reader field at the same time. An increase in tag density can lower the read robustness and hence the readability of a tag.

- **Tag motion.** The speed with which the tagged item is moving can impact tag readability because a tag needs to be present in the read window for a sufficiently long time to energize itself and/or transmit its data before moving out of it. If the tag moves out of the read window before it can successfully energize itself or has insufficient time to transmit its data properly, tag readability is affected. In these situations, you could use semi-active tags to neutralize the impact of speed on tag readability. Note that not all semi-active tags offer this benefit. It is important to note that almost certainly, at some point in the operations, the application needs to read the tag data when the tagged object is in motion (for example, during a loading or unloading phase). Therefore, a design assumption that completely rules out tag motion is most probably incorrect.

- **Tag orientation.** The way a tag is presented to the reader antenna plays an important role in the proper reading of this tag. Tag orientation depends on the antenna type and orientation. For a linearly polarized antenna, the tag has to be oriented to the antenna so that it can properly align itself with the electromagnetic field of the antenna. This means that if the linearly polarized antenna is oriented horizontally, the tag also needs to be horizontally oriented. Similarly, for a vertically oriented linearly polarized antenna, the tag needs to be oriented vertically. For a circular polarized antenna, you can orient the tag in almost any manner. You might recall from Chapter 1 that there is a common tag orientation—indirect orientation, for which the tag cannot be read (or read with difficulty) by any of the antenna types.

- **Operating environment.** This is discussed in Section 9.2.6, "Operating Conditions."

Tag readability varies with the type of the item to be tagged. It is difficult, if not impossible, to theoretically predict the tag readability correctly. Tag readability can *only* be reliably determined by setting up test RFID systems in the actual operating environment using actual RFID tags, readers, antennas, and the items to be tagged. It is important to do readability testing in the actual operating environment during normal operating hours, so that the testing can properly take into account all the variables that might bear on the readability. Such a setup might prove impossible because of various reasons. In this case, the testing should be done in a setup that mimics the actual operating environment as closely as possible. Even then, it is entirely possible that the readability results from such an environment will differ from the readability realized in the actual operating environment. Today, several vendors and integrators have set up their own RFID labs where they can perform readability testing on behalf of a business.

9.2.2.3 Data Capacity

Data capacity is a crucial variable that can have a great impact on application aspects such as cost and efficiency. The main questions that you need to ask here are as follows:

- How many data bits need to be stored on the tag?
- Is data lock needed?

The answer to the first question depends on whether the tag data is "license plate" type or contains other information, such as tagged object attributes. In the first case, most probably a 96-bit EPC Class 0+/1 or even a Class 0 tag might be fine. Why might? Because a designer might want to use a non-EPC type unique identifier, for example, which can be longer than a 96-bit EPC code. This practice is not totally uncommon in today's applications. In this case, the unique identifier actually encodes some object-specific data such as the expiry date, batch number, and so forth in it. Thus, it serves two purposes: First, the unique identifier is used to uniquely identify the item; second, a customer or operations personnel can use the same identifier to get some fundamental information about the item without having to access the enterprise product database. At the same time, the identifier is not so large (this is typically kept under 128 bits) so as to cause concern about the errors and delays associated with read transmission and write operations. If some specific tagged object information needs to be stored on the tag together with its unique identifier, a larger memory size is needed. Some examples of such tagged object data are its part number, maintenance history, and so on. It is not uncommon to use tags whose memory size is larger than 256 bits, and prototypical systems have successfully used tags with as many as 40,000 bits.

You might need data locks to prevent tampering with the tag data. You can lock a portion of the tag data so that it cannot be modified. A data lock can either be at the hardware level or at the software level. For Class 0 EPC tags, the unique identifier is burned onto the tag itself and hence is permanently locked from being changed. In contrast, a software lock (for example, for Class 1 EPC tags) can use a password scheme to unlock the locked data segments.

9.2.2.4   Physical Properties

This is another important variable that you should consider when designing an RFID system. Major concerns regarding physical properties include the following:

- **What are the right tag dimensions?** The proper tag dimensions generally depend on the size and shape of the tagged item. The dimensions might also depend on the real-estate size available on the product packaging, so that the item can be tagged without obstructing any critical product information. The proper physical tag size can be a lot smaller than the actual item size.

- **Does the tag have to be rugged?** This really depends on the environmental conditions under which the RFID system has to operate. The environmental conditions might include heat, cold, humidity, corrosive chemicals, mechanical shock, and vibration. A tag can be ruggedized to sustain some of these conditions. However, the price of such a tag might be more compared to the nonruggedized ones.

A tag should *never* be hacked by unwary individuals to bring the tag's physical dimensions down to an acceptable size for tagging an item. This includes folding the tag, cutting it using a pair of scissors, and so forth, which can seriously detune the tag antennas. If someone does so, modified tag might fail to draw sufficient power from the reader antenna signals, resulting in partial or total failure to transmit its data. Note that it is not uncommon to physically modify a tag (for example, drill holes into the tag antennas to increase its readability)! However, this should only be attempted by appropriate persons (such as the tag designers, for example) who know what they are doing.

9.2.2.5 Attachment

The way a tag is attached to an object plays an important role in the proper reading of the tag. This also varies with the item type. The following two questions must be answered in this regard:

- Where is the tag going to be attached?
- How is a tag going to be attached?

It is difficult to theoretically answer both of these questions. Experimentation is necessary in a test setup with the actual objects to be tagged and appropriate RFID hardware. Also, it is desirable that this activity be performed in the actual operating environment during normal operating hours so as to correctly factor in the variables that can affect the readability of the tag. This activity can be done in conjunction with the tag readability evaluation.

9.2.2.5.1 Where Is the Tag Going to Be Attached?

The tag should be attached to the object so that it does not hide important information, such as product brand and so forth, on the product packaging. Therefore, there may be only specific areas on the product packaging where the tag can be placed. It is important to identify these areas early in the design phase. A good rule of thumb is to place the tag next to the UPC label of the tagged object, where operations personnel can easily locate it. If it is determined that a tag cannot be satisfactorily placed anywhere on an item packaging, the packaging layout might have to be changed! Although this might seem a little too extreme, remember that designing an RFID application is a two-way street. Therefore, the business cannot expect benefits from the technology without providing some flexibility of its own. In general, you should not attach a tag on crush or fold points (where it would be prone to damage).

9.2.2.5.2 How Is a Tag Going to Be Attached?

Generally, if the product is made of RF-lucent material, a simple adhesive back strip is sufficient to mount the tag on the desired object. The situation is complicated when the tagged object material is made of RF-opaque or RF-absorbent material. In these cases, a spacer made of some type of RF-friendly material, such as a foam board of a certain thickness, can be used to mount the tag on the object. The thickness of such a spacer is typically equal or proportional to one quarter of the frequency wavelength used. This generally works for objects made of metals or objects that contain RF-absorbent liquids. In case of metals, you can physically attach the tag antenna with the metal portion of the object to increase the effective tag dimensions. A tag can also be attached so that a part of it extends out of the item's metal portion. In case of liquids, you can use multiple reader antennas so that the resulting strength of the RF-signal reaching the tag is sufficient to energize it in spite of getting attenuated to some degree. (This means that the readers are transmitting synchronously and in-phase.)

9.2.2.6 Volume

The volume variable is concerned with how many tags need to be used to tag the objects in the application for a particular period of time (a year, for instance). If tags are recycled, this number can be reduced substantially. If tags are not recycled, this will be a recurring overhead to operate the RFID application. When ordering tags, take care to order sufficient spares (depending on the error rate of tag creation as well as damaged tags). Currently, faulty tag rates can be as high as 20 percent or more!

9.2.3   Readers

Figure 9-6 provides a sample reader template.

Variable Name	Reader		
Selection Summary			
Constituting Variables			
Features	Type	Frequency	
		Embedded Controller?	☐ Yes ☐ No: Description
		Antenna Ports	☐ 2 ☐ 4 ☐ Other _____
		I/O Ports	
		Serial?	☐ Yes ☐ No
		Network?	☐ Yes: ☐ Wired ☐ Wireless ☐ No
		Stationary?	☐ Yes: ☐ Printer ☐ No
		Handheld	☐ Yes ☐ No
	Regulatory Information	Power	
		Duty Cycle	
	Upgradeability	☐ Yes: Description _____ ☐ No	
	Maintainability	☐ Yes: Description _____ ☐ No	
	Ruggedness Required?	☐ Yes: Description _____ ☐ No	
Installation			
Volume			
Notes			

Figure 9-6   Reader template.

This variable is further composed of the following subvariables:

- Features
- Installation
- Volume

A reader occupies the central position from the perspective of data collection. Therefore, much of the system's data-collection capabilities directly depend on the readers used. The new breed of readers provide sophisticated features such as SNMP (*Simple Network Management Protocol*) for real-time monitoring and management in a networked environment, data aggregation, and so on, which underlines the central role played by the readers in an RFID system.

The following subsections discuss each of these subvariables.

9.2.3.1   Features

The following main factors are to be considered in this context:

- Type
- Legal restrictions
- Upgradeability
- Manageability
- Ruggedness

As you can understand from this list, you must consider several reader factors so that it can successfully meet the application requirements.

9.2.3.1.1   Type

First, the reader has to operate on the selected frequency. If more than one frequency type is needed (for deployment in different geographical regions with varying frequency restrictions), it would be perfect if the reader can support these in the form of a multiple-frequency reader. Second, the reader needs to support at least two antenna ports (although four ports is becoming the norm). It is generally a safe bet to use a four-port reader because these offer better read zone coverage without a substantial price difference compared to a two-port reader. Third, the reader *might* need to support the number and type of I/O ports needed by the application. If a suitable controller is used that can provide these I/O ports, this is not required from the reader. Fourth, the reader controller is needed to support interfaces for sending tag data sets, I/O control, monitoring, and management. A controller is generally offered as a part of standard feature sets of almost any reader available today. However, depending on the application requirements, some specialization of these features might be necessary (such as support for a particular I/O port type for a particular I/O device [for example, a motion sensor]). However, such a reader might not even exist. Fifth, does the application require serial readers or network readers? Serial readers are not susceptible to network failure because they connect to the host computers over a serial connection. In contrast, network readers are flexible to configure as network devices, leading to better remote management, and they typically do not need as many host computers as needed by the serial read-ers. Sixth, the application might require stationary or handheld readers. Handheld reader area is currently undergoing rapid improvements. Therefore, a handheld reader matching your applica-tion requirements today might not be available yet. Take proper care when using a handheld reader to read a single tag. If more than one tag is present in the vicinity, a handheld reader might read several tags at once, including the tag of interest. This situation can also happen with station-ary readers. For stationary readers, you can alleviate issue by using *attenuators* (see Chapter 1). Although the wired readers are most frequently used, depending on the application requirements the reader might need to be wireless. The wireless requirement is generally targeted to a handheld reader.

9.2.3.1.2   Legal Restrictions

Most countries have legal restrictions on the transmitter power and duty cycle of a reader (see Chapter 1). Therefore, you need to confirm compliance of your selected reader type(s) with these limitations. Also, a reader's transmit cycle power should not be increased (decreasing it is fine) without express permission from regulatory bodies. Therefore, you should not tamper with a reader that is already certified as compliant with the legal restrictions. Doing so might violate the regulatory body license(s) and void the reader warranty.

9.2.3.1.3   Upgradeability

To maximize your hardware investment, a reader's firmware should be upgradeable for future enhancements and for fixing existing bugs in the current firmware. This can lead to a consider-able amount of cost savings when the existing application is upgraded or maintained at its current capabilities. It is strongly recommended that any reader, either stationary or handheld, selected for an application should be upgradeable. The capability of remotely upgrading a reader is prefer-able so that upgrades can be performed centrally in an optimized manner. A centralized upgrade eliminates the personnel time required to visit the installation site and manually upgrade each reader in turn.

9.2.3.1.4   Manageability

You want a reader that you can manage remotely so that you can centralize and automate moni-toring and management of reader's health. In case of a reader malfunction, you can remotely (and efficiently) track, diagnose, and fix the error. With such readers, you have minimum need for any service personnel to visit the installation site and prolong the operation downtime.

9.2.3.1.5   Ruggedness

Generally, a reader is not built as a rugged device. Depending on the requirements, however, a reader can be ruggedized. For example, a stationary reader can be enclosed in a rugged housing to prevent it from being affected by extreme cold, shock, vibration, and so on. A reader vendor can customize a reader to meet specific ruggedness requirements.

9.2.3.2 Installation

Although selecting the right reader is important, proper installation of it also requires good deal of attention, despite its mundane nature. Proper installation of a reader might require additional structures such as a portal to be built at the read zone. The reader and its cables cannot dangle in a manner such that it poses a risk to the operations personnel. Also, a reader might need to be fine-tuned (for example, by attaching in-line attenuators to the antenna ports to focus its transmitted energy in a smaller area and minimize interference). Long cable extensions connecting a reader with an antenna attenuate reader signals. Therefore, if the distance between the reader and the antenna is more than 6 feet (approximately 2 meters), you might need to use a high-quality low-loss cable. However, these types of cables are expensive and could increase the installation cost. The physical layout of the environment surrounding the reader antennas can also require a significant amount of installation effort. In short, proper installation of readers represents a nontrivial task that needs to be handled by qualified professionals.

9.2.3.3 Volume

This variable deals with determining the number of readers the application needs. This number depends on the number of read zones in the application, the number of readers needed to adequately cover each read zone, and the number of antenna ports supported by a reader. In general, the number of readers needed, each with four antenna ports, will be less than the number of readers needed if they have two antenna ports each. However, this can be complicated by the fact that a particular read zone setup might be such that only two antennas can be used per reader to avoid reader interference. Spare readers need to be considered as well to prevent operations interruption in case of a reader failure.

9.2.4 Antennas

Figure 9-7 provides the sample template for the antenna variable.

Variable Name	Antenna		
Selection Summary			
Constituting Variables			
Type	☐ Linear polarized	☐ Circular polarized	
Footprint			
Regulatory Information	Power		
	Duty Cycle		
Installation			
Volume			
Notes			

Figure 9-7   Antenna template.

This is another significant variable when you are designing and implementing an RFID system. This variable consists of the following subvariables:

- Type
- Footprint
- Power and duty cycle
- Installation
- Volume

These are among the simpler variable types and are relatively easy to tackle. This variable is easier to deal with because of the limited options available with regard to reader antennas in terms of features and complexity.

9.2.4.1   Type

You might recall from Chapter 1 that there are two types of antenna: circular polarized and linear polarized, if UHF is used. If the application requires that tags can be arbitrarily oriented to the antenna, you should use a circular polarization antenna. If a longer read range is desired or if the tag orientation to the antenna is fixed, you should use a linear antenna used. Antennas are generally not rugged, but can be ruggedized by the vendor.

9.2.4.2   Footprint

The antenna footprint, as discussed in Chapter 1, is a three-dimensional region consisting of RF waves emanating from the reader. A tag, when placed inside this region, can be read by the reader attached to the antenna. In reality, antenna footprint shape is rarely symmetrical because it is affected by RF interference in the operating environment. To determine the actual footprint map precisely, you need specialized equipment such as a spectrum analyzer or network analyzer. The hardware vendor or hardware services consulting companies can help with mapping the antenna footprints at the operations site.

9.2.4.3   Power and Duty Cycle

As detailed in Chapter 1, reader antenna power and duty cycle are constrained by regulatory bodies. Antenna power and duty cycle cannot be increased without express permission from the appropriate authorities, but you can always decrease it (useful in the case of antenna power if you need to limit the read distance). A decrease might prevent the RFID system from interfering with any neighboring RFID systems or reading tags that are out of range (for example, in an access control type of application). Power reduction can seriously affect the antenna footprint.

9.2.4.4   Installation

You should install an antenna as close to tags as possible and as far away from metals as possible (in UHF and microwave) to reduce reflection. FCC requirements for installing reader antennas states that an antenna should be positioned in such a manner that the personnel in the area for prolonged periods may safely remain at least 9 inches (about 23 centimeters) in an uncontrolled environment from the antenna's surface.[1] Antennas attached to different readers can cause reader interference when their read zones overlap. Therefore, you must take care to avoid this issue. Antennas might need to be installed in various positions to maximize the tag readability (for example, on the sides of dock doors, below a moving conveyor, above a canopy). You might have to construct specialized structures for this purpose. To maximize overall coverage area, you might have to tilt an antenna or install it at a particular height. Such tweaking depends on factors specific to where the antenna needs to be installed (such as dock door dimensions and so forth).

Consider the following regarding antenna installation:

- Interference
- Tag location
- Tag density
- Tag motion
- Tag orientation

---

[1]   FCC OET Bulletin 56. "Questions and Answers About Biological Effects and Potential Hazards of Radio Frequency Electromagnetic Fields." Bulletin 65, "Evaluating Compliance with FCC Guidelines for Human Exposure to Radiofrequency Electromagnetic Fields."

The last three variables have already been discussed. The following subsections discuss the first two variables.

9.2.4.4.1   Interference

This includes reader interference, which has already been discussed. Also, to avoid human interference issues, you might need to install antennas at particular vantage points in the read zone. Some older wireless LANs in the 900 MHz range can interfere with the readers.

You can mitigate the interference to some extent by placing reflectors (such as sheets of Mylar) in the packaging to help bounce the RF signal to the reader.

Interference is also discussed in Section 9.2.6, "Operating Conditions."

9.2.4.5   Tag Location

You must determine the coordinates (position and height) of where a tag is going to be located in the read zone to determine where to mount antennas. If the tag location range is large, you will probably need multiple antennas to read all the tags in the read zone. You might need to mount antennas on the sides and on the top and or the bottom of the read zone. To attain longer read ranges, you should use linear polarized antennas.

9.2.4.6   Volume

The number of antennas needed for an RFID application depends on the number of read zones supported by this application and the number of antennas needed to cover each read zone adequately. Think about spare antennas when determining how many antennas you need. Spares will help prevent operations impact in case of an antenna failure.

9.2.5   Items to be Tagged

Figure 9-8 provides a template for this variable.

Variable Name	Item		
Item ID			
Constituting Variables			
Type	☐ Pallet	☐ Case	☐ Item
Material			
Packaging			
Handling			
Notes			

Figure 9-8   Item template.

The importance of this variable cannot be overstated. In short, how well the items of an RFID application can be tagged plays a significant role in determining its success or failure. The following factors need close attention from the designers:

- Type
- Material
- Packaging
- Handling
- Speed
- Orientation to the antenna

The last two factors have already been discussed previously in this chapter. The following subsections analyze the impact of the remaining subvariables on an RFID application design.

9.2.5.1 Type

The item types can be broadly divided into the following three categories:

- **Pallet.** This item type is generally the easiest to tag and stands at the bottom of the complexity scale. However, certain pallet types (for example, pallets made from yellow pine) absorb moisture and attenuate RFID signal considerably and thus lower tag readability. Note that pallet tagging deals with item aggregates (cases of items) and, as a result, does not raise privacy-rights concerns.
- **Case.** This item type is slightly more difficult to tag compared with the pallet. Again, because case tagging deals with a collection of items, it is not generally associated with any privacy-rights infringement issues. However, in some stores where a customer buys items in cases (in bulk), this could be an issue.
- **Individual product.** This item type is generally the most difficult to tag. It can also involve privacy-rights issues (see Chapter 5, "Privacy Concerns").

Note that depending on the reading requirements, the relative degree of difficulty in reading a pallet, case, and individual product can vary.

9.2.5.2 Material

If the item is made of RF-lucent material for the frequency used (such as paper, dry wood, certain types of plastics, and so on), tagging such an item is much easier compared to situations where the item material type is RF-opaque (metals) or RF-absorbent (water, shampoo, and so forth).

9.2.5.3 Packaging

Packaging material plays an important role in determining tag readability of an item. If the packaging material contains metal foil, conductive carbon, or graphite, it might become difficult to properly read the tag attached to this packaging type. Also, a packaging material might absorb moisture to such an extent that it might act as an RF-absorbent material, leading to low tag read-ability. In these cases, the manufacturer might need to change the packaging material to strike a balance between the product advertisement needs and proper tag readability. If an item consists of multiple constituent objects (for instance, an eight AA battery pack), tag readability of this item might also depend on how densely the individual constituent objects are packed inside this item. In addition, for case-level tagging on a pallet, if the cases are packed too deep (depending on the case size and the packing depth), the case tags located near the center of the pallet might not be read at all (or with great difficulty), even assuming the use of an optimally tuned RFID sys-tem (irrespective of whether the material type of the case is RF-friendly). The reason is that the RF energy can penetrate only up to a certain depth, even in the case of RF-friendly material, depending on the output power and duty cycle of the readers. If an RFID tag is located at or below this depth, the RF energy will not be able to reach this tag. As a result, this tag cannot be read. Also, if pallet cases are small and densely packed, the antennas of some of these tags might cou-ple the antennas of other tags (for example, by physically touching one another). (Note that this is frequency dependent and antennas do *not* have to physically touch each other to couple and detune.) As a result, these tags cannot draw power from the reader antenna and hence cannot be read. This is known as the *shadowing effect*. The items should be packed so that the shadowing effect is minimized.

9.2.5.4   Handling

The methods and equipment used for handling a tagged object can impact tag readability. For example, if forklifts are used, the metal forks and the chain-lift assembly can reflect and prevent RF waves from reaching the tags, which are located closest to these handling parts. In addition, the wireless communication devices used by forklift operators might cause RF interference and impact readability. Last but not the least, the speed at which the forklift is moving can also affect readability. The effect of speed on readability has already been discussed. Similarly, for any other equipment type, the presence of metal, RF interference, mechanical vibration, *electrostatic dis-charge* (ESD), and motion can all impact tag readability.

9.2.6   Operating Conditions

Figure 9-9 provides a sample template for operating conditions.

Variable Name	Operating Conditions	
Summary		
Constituting Variables		
Type		
Footprint		
Regulatory Information	Power	
	Duty Cycle	
RF Obstructions		
☐ No		
☐ Yes		
Description: _____		
Environmental Conditions		
☐ Extreme Heat: Description_____		
☐ Extreme Cold: Description_____		
☐ Moisture: Description_____		
☐ Shock: Description_____		
☐ Vibration: Description_____		
☐ Static: Description_____		
RF Interference		
☐ No		
☐ Yes		
Description: _____		
Notes		

Figure 9-9   Operating Conditions template.

No two operating conditions are identical, even if they seem so. Operating conditions are like a person's fingerprint—even if two individuals look eerily similar, their fingerprints will not. Operating conditions can have a substantial bearing on the proper working of any RFID application. These conditions can be broadly classified as follows:

- RF obstructions
- Environmental conditions
- RF interference

It is difficult if not impossible to theoretically determine the impact of these variables. An RFID system needs to be tested at the actual operating site during normal operating time to determine the sum total impact of all the environmental condition variables on the system. Remember that a high-performance RFID solution designed for a particular business's operating conditions might perform poorly in its customers', distributors', or trading partners' operating environment. Therefore, designers and implementers need to consider these operating conditions when developing such an RFID application.

9.2.6.1   RF Obstructions

RF obstructions might result from the presence of RF-opaque objects and mobile equipment that effectively blocks RF waves from reader antennas reaching the tags. In addition, there could be RF-opaque objects, such as metal, in the environment that reflect the RF waves with similar impact. Finally, the environment might also contain dampeners or attenuators, such as people, RF-absorbent liquids, and construction materials, that can weaken the RF waves considerably, leading to read failures. In fact, the presence of human traffic is probably a certainty in any operating environment and can present a substantial challenge to overcome. The impact of these factors varies with the frequency used. The effect of these factors is most pronounced in UHF and microwave frequencies, and least in the LF and HF.

9.2.6.2   Environmental Conditions

Environment conditions include the presence of static, moisture, extreme heat or cold, shock, and vibration, all of which can affect the proper working of tags and shorten the life of tags, readers, and antennas. In addition, the presence of a substantial amount of moisture in the environment can dampen RF signals, resulting in insufficient energy reaching the tags for initiating data transfer. As a result, invalid reads and read failures might occur.

9.2.6.3   RF Interference

An existing installation base of RFID systems, which might or might not use the same frequency and type (for example, backscatter or transmitter) as the target system, might interfere with the latter. In addition, RF devices such as wireless radios, Wi-Fi, and Bluetooth can significantly contribute to the interference problem, too. Also, conveyor motors and controllers can generate RF noise that can interfere with the RFID system (but not to a substantial degree).

9.2.7   Vendors

Figure 9-10 provides a sample template for capturing vendor details. Figure 9-11 provides an example test case template.

Variable Name	Vendor		
Name			
Address			
Vendor Type	☐ Tag ☐ Reader ☐ Antenna ☐ Integrator ☐ Other: Description _____.		
Contact	Name		
	Phone		
	Fax		
	E-mail		
	Position		
Years in Operation			
Product	Name		
	Product Code		
	Features		
	Issues		
	Client Installations		
	Name		
	Product Code		
	Features		
	Issues		
	Client Installations		
Total Score on Test Cases (%)			
Notes			

Figure 9-10   Vendor template.

Test Case Name	
ID	
Description	
Steps	
Expected Outcome	
Result	
<Vendor Name>	Actual Outcome
	Score (%)	
	Comment
<Vendor Name>	Actual Outcome
	Score (%)	
	Comment
Notes	

Figure 9-11   Test Case template.

The data-collection capabilities of an RFID system is as good as the RFID hardware (that is, the tags, antennas, and readers), the RFID software used by the system, and how these components are tied together by the integrator(s). For example, although the RFID hardware currently offered by vendors is largely comparable, some vendors can offer specific hardware features that almost perfectly match the requirements of the RFID application under consideration. Therefore, selecting the right vendor(s) can be a crucial part of the application design and determine the degree of success of the system. Keep in mind the following factors when selecting system vendor(s):

- Use actual evaluation
- Avoid vendor lock-in
- Select multiple vendors
- Plan for upgrade
- Vendor support
- Vendor relationship
- Plan for contingency

Vendor selection activity can commence as soon as the application (functional and non-functional) requirements are finalized. Vendor selection should not be done in haste. The test cases for evaluating the vendors should be carefully crafted based on actual application requirements. Then the vendors should be invited to participate in the evaluation effort (or a test lab can do this on behalf of the business using different vendor hardware). Vendors should then be evaluated and ranked based on how they fared on the test cases. The top three vendors may now be considered as potential candidates and can further be narrowed down based on their references, financial stability, and so forth. This entire effort from designing test cases to the selection of the winning vendor can easily take weeks. So this activity needs to be planned ahead and executed as early as possible in the design phase.

The following subsections discuss each of these variables.

9.2.7.1 Use Actual Evaluation

Vendor selection should be based on actual evaluation results. It is not prudent to select a vendor based on its hardware or software literature and data from case studies and "reference cases." Just because the hardware or the software worked for another application, which might even be similar, you have no guarantee that it will work for the application in question. The actual evaluation of the hardware using pertinent test cases should be performed, most preferably at the actual operation site, during normal operating hours. The vendor(s) should be able to prove that their hardware really works for the actual items under the actual operating conditions.

9.2.7.2 Avoid Vendor Lock-In

One of the best strategies to bypass vendor lock-in is to avoid proprietary solutions from a vendor and base the RFID application design on a well-supported standard. Hardware and software from the different vendors that support a particular standard have a higher chance of being compatible with each other than the proprietary solutions from these vendors.

9.2.7.3 Select Multiple Vendors

If you cannot find a single vendor that can perform end-to-end implementation, it is fine to select multiple vendors, some of which might supply the tags, some the readers and antennas, and the remaining ones the software and integrating service. However, the hardware and software should be compatible with each other so that the application can be implemented by integrating these components together. The best way to guarantee this is to base the application design on a well-supported standard and use the compatible components from the vendors.

9.2.7.4 Plan for Upgrade

RFID technology is undergoing rapid changes, and vendors are bringing out new hardware, firmware, and software upgrades at a furious rate. To prevent hardware and software obsolescence and continuous upgrading effort, designers need to come up with an upgrade strategy. This strategy can include items such as using backward-compatible hardware and software from the vendors, defining minimum upgrade intervals, ability to upgrade existing reader firmware, and so forth. However, *at present, upgrading to new hardware seems like an unavoidable option.* So this must be planned for properly in advance.

9.2.7.5   Vendor Support

If a vendor cannot support its hardware or software in a timely fashion in the deployment region(s), it is debatable whether it can be a preferred vendor. Never assume, for any reason, that support at any point is unnecessary. On the contrary, support is always a necessity from the hardware, as well as from software and possibly from services, perspective and should be built in to the plan accordingly.

9.2.7.6   Vendor Relationship

The importance of having a good vendor relationship can hardly be overstated. This might not seem obvious at first. After all, doesn't a vendor relationship end with the timely delivery and maintenance of its products? Not really. As RFID technology is maturing, the vendor products are also changing at a rapid pace. *The business should take this opportunity to provide the vendor with guidance to improve its product line.* For example, the business can use the results from its RFID pilot to report issues with tag readability and the conditions under which it is happening. This information will allow the vendor to rectify the issues and present the business an improved version of its product. This symbiosis will benefit the business, the vendor, and the RFID technology in the long run. Also, if special feature support is required by a business application but is currently missing, the vendor can deliver a prototype version of its hardware that includes this feature. This prototype will allow the business to meet its delivery timeline without the need for the vendor to upgrade its product line overnight (sacrificing its delivery and quality assurance processes). This type of cooperation is greatly facilitated by a good relationship between the business and its vendor. Therefore, *a vendor should not be treated as an adversary, but as a partner with whom the business vision should be shared.* After all, you want the vendor to understand and anticipate the business needs and align its products accordingly.

9.2.7.7   Plan for Contingency

What happens if there are issues with the selected RFID vendor down the road? Among other things, a vendor might not be able to supply the hardware or software on time, balk at the maintenance promises, go bankrupt, encounter patent-infringement lawsuits (and can then go bankrupt as a result), or simply decide that it might not want to play in the RFID arena anymore. Designers need to have a contingency plan to address these issues. Again, if the design is based on a widely accepted standard, you might be able to replace one vendor's products with those from another vendor (that also supports the selected standard) without causing a major catastrophe in the application delivery.

9.2.8 Standards

Figure 9-12 provides an example Standards template.

Variable Name	Standard		
Summary			
Constituting Variables			
Organization			
Title			
Version			
Description			
Notes			

Figure 9-12   Standards template.

Any RFID application design should be based on a standard for the following reasons:

- Savings of time and effort
- Collaboration
- Scalability
- Freedom of choice

These are critical success factors for an RFID system (or any system, for that matter) in the long run. It is tempting to craft a proprietary system in the haste of pushing out the first version of an RFID application in the field. This is especially true for a closed-loop system for which there is no need to share the data with outside world. Although such an effort might be successful in the short run, the chances are high that it will run into a number of issues in the long run, particularly when the need to share data with external entities arises. Therefore, the recommendation is to resist such a temptation and invest some time upfront to investigate whether a suitable standard exists that can be leveraged for the solution. The following subsections discuss these advantages of standardization.

9.2.8.1 Savings of Time and Effort

A standard typically provides a ready-made solution for a set of business problems. By basing the solution design on a standard, the designer is saving himself the time and effort associated with a long and often frustrating effort of building such a solution from scratch. Indeed, if a proprietary solution is used, the designers must provide good reasons for foregoing a standard-based solution.

9.2.8.2   Collaboration

An RFID system has to be compatible not only with the internal systems of the business that has implemented it, but also with external systems such as the systems of suppliers, customers, and business partners. The maximum potential of an RFID solution is usually realized by collaborating with these external entities, not with the entities bound within the four walls of an enterprise. A proprietary solution might have a limited prospect of success regarding compatibility with these third-party independent systems. However, a system based on standards offers the best solution to achieve this goal.

9.2.8.3   Scalability

Scalability is an important aspect of virtually any system that can grow with the business needs. A standard-based solution offers the best prospect for scalability because this aspect is typically built in to the standard itself. However, for a custom solution, scalability might be difficult to achieve because of shortcomings in the solution that were not apparent in the early phases.

9.2.8.4   Freedom of Choice

If a standard-based solution is used, the solution components can be mixed and matched from different vendors that support the standard. As a result, the business is not locked-in to any particular vendor. This flexibility is desirable for a technology such as RFID, which is currently dominated by small players. Such a company might not be able to survive unfortunate events such as a legal suit bought by a competitor and so on. It is the designer's responsibility to take these issues into consideration and insulate the business and the application from such impacts. In contrast, a proprietary solution has a good chance to become dependent on a single vendor or a handful of vendors.

9.2.9   Application Software and Hardware

Figures 9-13 and 9-14 provide sample Software and Hardware templates, respectively.

Variable Name	Software		
Summary			
Constituting Variables			
Controller	☐ No ☐ Yes: Description _____		
	Volume		
Middleware	☐ Build ☐ Buy		
	Volume		
	Description		
Custom Application	☐ Build ☐ Buy		
	Volume		
	Description		
Database	Name		
	Volume		
Operating System	Name		
	Volume		
High Availability	☐ No ☐ Yes: Description _____		
Notes			

Figure 9-13   Software template.

Variable Name	Hardware	
Summary		
Constituting Variables		
Host Machines	Type	☐ Personal Computer ☐ Server
	Volume	
	Configuration	
	Type	☐ Personal Computer ☐ Server
	Volume	
	Configuration	
Controller	☐ No ☐ Yes: Description _____	
	Volume	
Annunciator	Type	
	Volume	
	Description	
	Type	
	Volume	
	Description	
Actuator	Type	
	Volume	
	Description	
	Type	
	Volume	
	Description	
Sensor	Type	
	Volume	
	Description	
	Type	
	Volume	
	Description	

Figure 9-14   Hardware template (continues).

Network Router	Type	☐ Wired ☐ Wireless	
	Volume		
	Description		
	Type	☐ Wired ☐ Wireless	
	Volume		
	Description		
High Availability	☐ No		
	☐ Yes: Description _____		
Notes			

Figure 9-14   Hardware template (continued).

Most of the general practices used for application software and hardware design hold for designing an RFID application, too. Effort should be made to determine whether any existing hardware and software assets can be reused to implement and deploy the RFID system. The following issues should be addressed, at a minimum, in this context:

- Data volume
- Software build versus buy
- High availability (HA) requirements
- Volume

The following subsections discuss these factors.

9.2.9.1   Data Volume

Massive amounts of data are typically generated by a nontrivial RFID system. This is partly because, unlike bar code, a tag is read multiple times when it is in the read window of a reader. This data needs to be filtered, aggregated to bring it down to a manageable amount when it can be persisted, and used. The RFID middleware should be responsible for handling this aspect. Hence, this software component plays a crucial role in data movement and management.

9.2.9.2   Software Build Versus Buy

This is a classic issue in any application design decision. The software component includes the RFID middleware and application-specific systems such as a fleet management system that the business might not currently own. If the design is based on a standard for which off-the-shelf implementations are available from various software vendors, it can be advantageous to buy this component, especially if the business does not have the appropriate in-house IT expertise and experience to develop this application and has tight resource and time constraints to develop and deploy this application.

9.2.9.3   High Availability (HA) Requirements

This option almost always will increase the implementation cost. If *continuous availability* is required, then besides additional hardware, the application/vendor software has to provide built-in support so that the database engine is available for transactions 100 percent of the time. This means two identical but fully independent systems are required, both in terms of software and hardware. The application software has to be aware of these two systems and treat every transaction as being distributed across the two systems. In short, this type of HA solution is the most complex and expensive and should be used for only the most critical systems. Fortunately, most of the RFID applications can use other types of HA. For example, the database engine, middleware, network adaptors, and so on can be tightly integrated with specialized HA software to produce what is known as an *HA cluster*. This can be configured in different ways. For example, in a *mutual takeover* option, each machine in the cluster is active and ready to take over in case of a failure; whereas in *hot standby*, the secondary machines are idle waiting for a failover. This is the next-most expensive solution after continuous availability. The cheapest failover solution involves *data replication*, which also offers the most limited form of failover. In this case, a standby system is used to back up data from the primary system. In case of failure, manual intervention is necessary to start the standby system to take over the responsibilities of the primary.

9.2.9.4   Volume

Volume is concerned with the quantitative details, such as the number of software licenses, annunciators, actuators, hardware servers, network routers, and so forth, needed to install and run the RFID application at the desired operating locations. With the exception of licensing and ongoing maintenance costs, this is generally a one-time cost.

9.2.10   Integration with Existing Systems

Figure 9-15 provides a sample Integration template.

Variable Name	Integration with Existing Systems		
Summary			
Constituting Variables			
Interfaces	Name		
	Implementation	☐ Build	☐ Buy
	Description		
	Name		
	Implementation	☐ Build	☐ Buy
	Description		
Notes			

Figure 9-15 Integration template.

Integration of an RFID application with existing systems might not be needed if the business does not have such existing systems; even when it does, the need for integration might not be desperate. The latter is especially true in RFID trials and pilots used for technology evaluation and business case validations. Also, if a company is just trying to design a slap-and-ship application (see Chapter 8, "Creating Business Justification for RFID") to achieve some sort of RFID compliance for its customers, it might want a bare-bones standalone RFID system with little or no integration with its existing business systems. This is so that the compliance is achieved while spending as few resources and impacting as few aspects of the operations as possible. In the long run, this practice should be avoided. An RFID system is a data-collection system; so if a business spends money to collect data but does not use it to its advantage, it is questionable how much benefit it is getting out of it. To realize the full potential of RFID, the enterprise should, at some point, be prepared to integrate the RFID application with its business processes and back-end system(s).

Some large enterprise systems (for example, ERP [*enterprise resource planning*] and WMS [*warehouse management systems*]) have already been RFID-enabled by their vendors. The business can look into using this capability if it is using one of these systems.

As a general guideline, if a custom interface needs to be developed, the temptation to implement the back-end functionality in the RFID middleware should be resisted. However, the business rules for data filtering and aggregation should be implemented at the middleware level because this is one of the principal functionalities of RFID middleware. But then the business-specific logic for using this processed data (that is, the business transactions) should be implemented at the back end. It is perfectly fine to *trigger* business transactions from the middleware and at interface levels, but *implementing* the business transaction at either of these levels is not. The latter implementation types might not scale and can cause synchronization issues in the long run with the back end in terms of logic and data (a maintenance nightmare).

9.2.11   Maintenance

Figure 9-16 provides a sample Maintenance template.

Variable Name	Maintenance		
Summary			
Constituting Variables			
Tag	Type		
	Volume		
	Type		
	Volume		
Damaged Hardware Replacement	Tag	Type	
		Volume	
		Type	
		Volume	
	Reader	Type	
		Volume	
		Type	
		Volume	
	Antenna	Type	
		Volume	
		Type	
		Volume	
	Controller	Type	
		Volume	
		Type	
		Volume	
	Annunciator	Type	
		Volume	
		Type	
		Volume	
	Actuator	Type	
		Volume	
		Type	
		Volume	
	Sensor	Type	
		Volume	
		Type	
		Volume	

Figure 9-16   Maintenance template (continues).

Hardware and Firmware Licenses	Name	
	Volume	
	Description	
	Name	
	Volume	
	Description	
Hardware and Firmware Upgrades	Name	
	Volume	
	Description	
	Name	
	Volume	
	Description	
Software Licenses	Name	
	Volume	
	Description	
	Name	
	Volume	
	Description	
Software Upgrades	Name	
	Volume	
	Description	
	Name	
	Volume	
	Description	

Personnel	Full Time	Employee	
		Consultant	
	Part Time	Employee	
		Consultant	
Technical Support	Software Contract		
	Hardware Contract		
	Infrastructure		

Figure 9-16   Maintenance template (continues).

Enhancements (Estimated Hours)	Custom Application	
	Middleware	
	Interface	
Notes		

Figure 9-16   Maintenance template (continued).

The maintenance activities and the required resources need to be planned in advance, not after the system has been delivered. A major part of TCO of an RFID system actually may lie in maintaining it. You can use the following as a checklist for determining maintenance parameters:

- **Tags.** Even if tags are recycled, new tags might be needed to tag the items as the system is in operation. This is because some tags can get damaged during the recycling stage. For applications that do not recycle tags, tag replacement is definitely needed.

- **Replacement of damaged hardware.** This includes RFID tags, readers/printers, antennas, and controllers. Usually not all these components are susceptible to damage equally. Tags probably need to be replaced most frequently, followed by antennas, then readers and controllers.

- **Hardware and firmware licenses (if applicable) and upgrades.** This involves licenses (if applicable) and upgrades for RFID readers/printers, antennas, and controllers. It also includes upgrades for personal computers, servers, and so on used for hosting the application.

- **Software licenses and upgrades.** This is an important parameter representing the licenses and upgrades for RFID middleware, custom modules, and applications from vendors, such as real-time inventory management, databases, operating system, reporting software, and so on.

- **Personnel.** This includes both part-time and full-time employees and consultants needed to maintain the system.

- **Technical support.** This encompasses resources for setting up a technical support environment, support contracts with the vendors and integrators, and so on.

- **Enhancements.** The customizations needed for the RFID system to support new functionalities or extend the scope of existing functionalities, new interfaces to back-end systems, and so forth are included in this parameter.

Some of these items, especially the cost of enhancements, can be hard to estimate accurately at design time.

9.3 Implementation Notes

This section makes several suggestions based on experience gathered from implementing real-world RFID systems. The suggestions are free-form and are listed here without any order of priority:

- Test, test, and test.
- Analysis first, hands-on next.
- Avoid scope creep.
- Need for multiple tag checkpoints.
- Composition relationships are important.
- Never overlook tag validation.
- Include expectation management.
- Remote monitoring and management capabilities are essential.
- Time process changes.
- Comply with regulations.
- Need for patience and hard work.
- Unexpected benefits.
- Involve smart people.

The following subsections explain what these mean.

9.3.1 Test, Test, and Test

An RFID system should be tested as thoroughly as possible, and then some. This is especially true in an iterative approach where the current system release has to be as bug-free as possible so that the next iteration can be built on top of it. Tag placements, antenna positions, and so forth all have to be tested rigorously until the system performs at a satisfactory level at its current scope. At that point, the system can be moved to the next iteration. Several vendors and associations have set up RFID laboratories that you can use for testing purposes. These labs lack real-world context, but provide opportunity to test tags without disrupting business operations.

9.3.2 Analysis First, Hands-On Next

It might come as a surprise to you how much information can be gathered by upfront analysis instead of blindly diving into the hands-on exercise instead. The important process flows and use cases should be analyzed as much as possible to glean the business characteristics and parameters before any hands-on activity begins. After the designers are confident that they have grasped the essentials, setting directions and strategy follows naturally, making the hands-on trials and projects much more focused and useful. Without this guidance, hands-on activities might dither and prolong unnecessarily. In one actual case, a business spent more than a year of hands-on experimentation without really coming to a conclusion to use (or not use) RFID!

9.3.3   Avoid Scope Creep

Laser-like focus should be maintained on the scope of an RFID system. It should not diverge to include features that were not originally included when the design and analysis work was done. It is easy to get carried away by the success of a throwaway pilot and be unrealistically optimistic about the technology prospects. For example, the scope might include changing business processes when the other parts of the system are poorly understood or analyzed. As a result, the system's scope, scale, and complexity might grow to a point that the analysis, design, and implementation encounters major roadblocks that jeopardize the delivery timeline. *Scope should be handled in iterations*. Scopes should be prioritized using some criteria (such as ROI, risk, and so on) and then iteratively implemented in such a manner that the total risk is always controllable for each such iteration.

9.3.4   Need for Multiple Tag Checkpoints

A reader validates a tag at the point of its creation. However, additional checkpoints are necessary, either manual or automatic, at other points in the operation. The reason is simple. A perfectly working tag might become damaged after it passes through a certain number of processing steps in the operations, rendering it completely useless. If a read of this tag is then attempted at any point, the read will fail. To catch this problem, multiple checks needs to be put in place in the operations to make sure that a tag is not damaged and can be properly read. If a damaged tag is found, corrective actions need to be taken to fix the problem (for example, creating a new tag with the unique identifier of the damaged tag and attaching it to the item). As you can understand from this discussion, merely slapping a tag on an item is no guarantee that it can be read at a later point in the operations. Therefore, even a slap-and-ship type application needs to validate tags after they are created and put on an item, most probably when items are leaving a business site (for example, when the item is being loaded onto a delivery truck for its customer).

9.3.5   Composition Relationships Are Important

The interrelationships among entities such as pallet, case, and unit are important. Therefore, a tag ID of a pallet (when loaded) should correspond to a set of case tags that this pallet contains, and each such case tag should then correspond to a set of unit tags that this case contains. Although unit-level tagging is not prevalent today, these are some examples of fundamental composition relationships that exist in business operations. Other composition relationships might also exist in a business. An RFID system should maintain these relationships instead of being "flat," where the associated tag data for a pallet, case, and a unit are implemented at the same level of hierarchy.

9.3.6   Never Overlook Tag Validation

It is easy to get hung up on tag readability and overlook an equally important aspect—tag validation. Tag validation means that the tag data is checked against some criteria to make sure that it is valid. This is generally done using business-specific logic and data stored on the back end in conjunction with the tag data. A tag whose readability is close to perfect but contains data that is invalid is not useful to a business. You would be surprised how often this attribute is sacrificed in favor of tag readability.

9.3.7   Include Expectation Management

Expectations need to be managed carefully in an RFID system implementation effort. The designers and implementers need to make the stakeholders aware of the benefits that can be realistically achieved from such a system. Business needs to be patient about expecting savings from a newly minted RFID system because this system might need time for fine-tuning to perform optimally. A slew of business practice eliminations or changes should *not* be undertaken with the assumption that an RFID application will replace or provide a superior alternative to these practices *before* actually validating that the application can deliver on its promises.

9.3.8   Remote Monitoring and Management Capabilities Are Essential

Remote monitoring does not seem to be a necessity until some system component fails and then business operation is impacted to troubleshoot and fix the failure. An RFID system might not be maintainable unless a monitoring capability is built in to the system. Similarly, if a business site has 200 installed readers, manually upgrading the reader firmware to its latest version will need a substantial amount of personnel time and business downtime. To make matters worse, some of the readers might not need the upgrade or might need a different version of the firmware, thus making any manual update procedure prone to error. Remote management is an effective answer to this type of issue. Therefore, designers and implementers need to pay special attention to include these features in the system, which will make the system efficient and economical in the long run.

9.3.9   Time Process Changes

Although it is true that the maximum potential of RFID can be realized through the existing business process changes, it is truer that these changes should not be introduced at the early stages of implementation. In other words, unless the RFID system has proved itself in the actual business operations, the business process changes should *not* be implemented. Why? Because if an RFID system does not deliver according to the business expectations or needs time for fine-tuning, it can be isolated without causing any impact on the existing business operations. After the RFID system has matured to a certain acceptable point, the business process changes can slowly be introduced.

9.3.10   Comply with Regulations

Introduction of RFID systems should not violate any federal or state regulations or job safety codes. The business can be subject to hefty fines and might have to assume substantial liability if it does not comply with such regulations. For example, some items (such as pharmaceuticals) might require you to follow strict government guidelines for tagging; the hardware installations might need to be done by properly trained professionals who are knowledgeable about these codes and regulations (which might also reduce hardware damage and system downtime).

9.3.11   Need for Patience and Hard Work

No single one-size-fits-all RFID solution exists. Likewise, you cannot just buy a plug-and-play solution and use it without any modifications. The situation is made more challenging by the fact that a similar solution that has worked for one business might perform badly for another. This means that implementing a successful RFID solution is not easy. The variables involved (as described throughout this chapter) need to be methodically analyzed and resolved to arrive at a system that satisfactorily meets its requirements. Therefore, patience and hard work are essential ingredients to successfully take on an RFID project.

9.3.12   Unexpected Benefits

A pilot or an evaluation effort might reveal unexpected advantages that were not comprehended at the planning or design time. Such benefits might manifest themselves in the form of automating parts of the business as a side effect of the scope of the pilot, substantially more productivity gains than originally expected, and so forth. Therefore, implementing a pilot can offer additional advantages besides technology evaluation and validation of business assumptions.

9.3.13   Involve Smart People

Designing and implementing a nontrivial and successful RFID application takes a lot of effort by bright people. If a business expects to engage second- and lower-rate talents just because these individuals are available, be forewarned. The early stages of RFID adoption should be in the hands of the people who not only have deep technical skills but also business domain knowledge. These people should be able to separate reality from hype, set realistic expectations with the stakeholders, and have unrelenting drive to apply and improve the system to exploit its optimum potential.

9.4   Conclusion

This chapter covered several aspects of an RFID solution from design and implementation perspectives. The influencing variables were classified and discussed in a manner so that you can understand the relationships of these variables with the different components of the RFID system. In this chapter, the variables were decomposed into their constituent subvariables to provide you an idea about the structure and composition of the variable. Several practical hints and suggestions were provided to help guide you to make the right decisions for your RFID application.

CHAPTER 10

Standards

A discussion of RFID technology is not complete without the standards from different standards bodies and organizations aimed at resolving and standardizing different aspects of the technology. Why should you be concerned about standards? For the following good reasons:

- **Design and implementation of a robust, well thought-out working system.** Instead of spending resources on crafting a proprietary system from scratch, which might be prone to errors and deficiencies, the appropriate standard might specify a solution that has undergone several iterations and improvements, which will enable you to produce a solution that is well-defined, robust, and tried and tested in real-world implementations.

- **Design and implementation of an open system.** This is another huge benefit of using standards. The standard can provide strict specifications for solution components that vendors and integrators provide off the shelf, and thus you avoid any lengthy development effort and vendor lock-in.

- **Design and implement a compatible system.** Another important benefit of using standards is that the resulting solution is compatible with a wide array of related systems. Therefore, you will require fewer resources and less effort to integrate this solution with other systems.

You are strongly advised to locate a standard or specification that matches your application requirements as a first step when designing and implementing a solution, *before making any effort at all* in building a proprietary solution. What happens if the application in hand is trivial or such a matching standard or specification does not exist? First, if the solution is trivial or throwaway (for example, a quick and dirty proof-of-concept type application), not using a standard could be justified if this solution is not going to be iterated to arrive at a deployable solution. Second, you can adopt a related existing standard; for example, you can apply the

EPCglobal specification (described later in this chapter) in almost any passive UHF RFID application. In the rare case that you can find no kind of standard to follow, you can still explore certain areas (such as tag types [frequency dependent] to use, how to attach RFID tags to items, and so on). You can still learn valuable lessons (instead of giving up completely) regardless of whether a standard is available.

Many RFID standards are available from many different organizations around the world. Descriptions of existing RFID standards could easily fill a book. Because of the changing nature of the standards, however, a static description in such a book would be outdated even before publication. Therefore, this book attempts a different approach. This chapter discusses RFID standards that are in existence today, but makes no claim that the standards covered here are exhaustive. For example, new standards might be adopted subsequent to this writing. Similarly, some standard might disappear totally or have parts withdrawn. If you are interested in a certain standard and want to explore it in detail, procure a copy of the actual specification by contacting the appropriate standards body. You can find contact information for these standards bodies at the end of this chapter. That said, this chapter does describe some select standards. However, these are not a substitute for the actual standards specifications. As stated previously, standards change and evolve over time; therefore, some of the description provided at the time of writing might have changed by the time of reading. The following major standards organizations, in no particular order, have either produced standards related to some aspect of RFID or have provided related regulatory functions:

- ANSI (American National Standards Institute)
- AIAG (Automotive Industry Action Group)
- EAN.UCC (European Article Numbering Association International, Uniform Code Council)
- EPCglobal
- ISO (International Organization for Standardization)
- CEN (Comité Européen Normalisation (European Committee for Standardization))
- ETSI (European Telecommunications Standards Institute)
- ERO (European Radiocommunications Office)
- UPU (Universal Postal Union)
- ASTM (American Society for Testing and Materials)

This is not an exhaustive list. The standards that appear in this chapter emanate from a subset of the preceding standards bodies. In addition, standards/mandates from government are discussed.

> **NOTE**
>
> If a standard is composed of multiple parts and information on some of the parts is not available at this time, the standard is assumed to consist of the remaining parts.

10.1 ANSI Standards

ANSI is a private, nonprofit organization that administers and coordinates the U.S. voluntary standardization and conformity assessment system. The institute's mission is to enhance both the global competitiveness of U.S. business and the U.S. quality of life by promoting and facilitating voluntary consensus standards and conformity assessment systems, and safeguarding their integrity.

Some major ANSI standards relevant to RFID technology and its use in real-world applications are listed here by title and with an optional brief description:

- **ANS INCITS 256-2001.** Standard for promoting interoperable RFID devices operating in freely available international bands and license-free power levels. It also supports item management applications.

- **ANS INCITS 371.** Information Technology—Real Time Locating Systems (RTLS). This is composed of the following three parts:
  Part 1. 2.4 GHz Air Interface Protocol
  Part 2. 433 MHz Air Interface Protocol
  Part 3. Application Programming Interface

- **ANS MH10.8.4.** ANSI application standard for RFID for reusable plastic containers. This is compatible with ISO 17364.

10.2 AIAG Standard

A nonprofit association, AIAG's primary goals are to reduce cost and complexity within the automotive supply chain and to improve speed to market, product quality, employee health and safety, and the environment.

The following RFID standard from AIAG enjoys widespread support from manufacturers and suppliers:

- **AIAG B-11.** Tire and Wheel Label and Radio Frequency Identification Standard. The current version provides a 96-bit number in EPCglobal (discussed later) data format for RFID tags and labels.

10.3   EAN*UCC Standard

EAN*UCC System is co-managed by the Uniform Code Council, Inc., and GS1 (formerly EAN International). The EAN*UCC System standardizes identification numbers, EDI transaction sets, XML schemas, and other supply-chain solutions that enable more efficient business processes.

The *Uniform Code Council* (UCC) (changing its name within the next few months to GS1 US) is a nonprofit organization dedicated to the development and implementation of standards-based, global supply-chain solutions. Under its auspices, the UCC operates three subsidiaries—UCCnet, RosettaNet, and EPCglobal US—and it co-manages the global EAN*UCC System with GS1. The UCC also manages the *United Nations Standard Products and Services Code* (UNSPSC) for the *United Nations Development Programme* (UNDP). EPCglobal, Inc., is a joint venture of the UCC and EAN International. UCC-based solutions, including business processes, XML standards, EDI transaction sets, and the bar code identification standards of the EAN*UCC System, are currently used by more than one million member companies worldwide.

The mission of GS1 and the member organizations is to create open, global, multisector standards based on best business practices, and by driving their implementation, play a leading role in supply and demand chain management improvement worldwide. GS1 has member organizations in 101 countries. The EAN*UCC System is an integral part of the way business is conducted worldwide. Today, more than one million member companies in 155 countries use EAN*UCC standards as part of their daily business communications, representing more than five billion scanning transactions a day. The supply-chain solutions offered by the EAN*UCC System include globally unique identification codes, data transport media, electronic commerce, and communications standards. These tools support established industries as well as emerging markets.

The following RFID standardization initiative is provided by this standards body:

- **GTAG (Global TAG).** This is aimed at facilitating global supply-chain operations in the 862–928 MHz (UHF) band. It provides a technical foundation with canonical data sets and applications guidelines. GTAG-compliant RFID tags are currently available from several vendors.

10.4   EPCglobal Specification

EPCglobal, Inc., is a joint venture between the UCC and EAN International. The aim of EPCglobal is to establish worldwide standards for designing, implementing, and adopting *Electronic Product Code* (EPC) and EPCglobal Network (described later). The EPCglobal specification (soon to be deemed a standard) targeted for supply-chain operations is probably the most promising global specification for RFID that can also be applied to a very wide array of applications.

A short history of EPCglobal is in order. EPCglobal, Inc., took over the administrative responsibilities of its predecessor Auto-ID Center on November 1, 2003. The research functions of Auto-ID Center were transferred to several worldwide Auto-ID labs. EPCglobal, Inc., maintains a very close relationship with Auto-ID labs to enhance the technology and meet future needs. Auto-ID Center was founded at M.I.T. in October 1999 as a partnership research program sponsored by 100 companies and 5 of the world's leading universities. It was responsible for conceptualizing, creating, and promoting the original specification called the *Auto-ID Center* specification that involved the EPC technology. Why was the transformation from Auto-ID Center to EPCglobal, Inc., needed? After the EPC technology was sufficiently developed in the research setting, the need of an experienced standards body was felt to commercialize and drive global adoption of the technology. Both EAN and UCC have several years of experience in handling standards, and the combination of these two bodies truly makes one of the most globally capable entities for advancing the EPC and EPCglobal Network.

The following section discusses EPCglobal Network, which is the fundamental component of the EPCglobal specification.

10.4.1   EPCglobal Network

The EPCglobal Network is a collection of technologies that can provide automatic, real-time identification and intelligent data sharing of an item both within and outside of an enterprise. Although this is geared toward the supply-chain operations of an enterprise, it can be applied in other types of applications (for instance, item tracking and tracing [see Chapter 4, "Application Areas"]), too.

Five main technology components make up the EPCglobal Network, as follows:

- Electronic Product Code (EPC).
- Data-collection hardware consisting of EPC tags and readers. This is also collectively known as *ID System*.
- EPCglobal middleware.
- Discovery Services (DS) composed of, for example, Object Naming Service (ONS).
- EPC Information Services (EPCIS).

Thus, the EPCglobal Network "equation" can be summarized as follows:

EPCglobal Network = ID System + EPC + Middleware + DS + EPCIS (1.1)

In addition, EPCglobal provides a reference architecture for the network.

The next subsections discuss these components in detail. These descriptions are followed by a section that explains how these components interact together to form the EPCglobal Network.

10.4.1.1   Electronic Product Code (EPC)

The *Electronic Product Code* (EPC) is a license-plate type identifier that can *uniquely* identify any item in a supply chain. It is a simple and compact scheme that can generate extremely large quantities of unique identifiers. At the same time, this scheme allows accommodation of legacy codes and standards such as the following:

- **Global Trade Identity Number (GTIN).** This is a globally unique EAN-UCC number for identifying products and services.

- **Global Returnable Asset Identifier (GRAI).** This is used for numbering returnable assets such as drums, gas cylinders, and so forth.

- **Unique Identification (UID).** This is a U.S. Department of Defense numbering scheme for asset tracking.

- **Global Location Number (GLN).** This is used for representing location, trading partners, and legal entities.

- **Global Individual Asset Identifier (GIAI).** This is used to identify immovable asset as well as fixed inventory of a business.

- **Serial Shipping Container Code (SSCC).** This is used to identify shipping units such as a pallet, case, carton, and so on.

A company that uses bar codes in its operation today can have a migration path to RFID using EPC. An EPC code can be used to determine various attributes of an item, such as the following:

- Version of the EPC used

- Manufacturer identification

- Product type

- Unique serial number of the item

Two EPCs can be of different sizes. Currently, 64 bit and 96 bit are the most predominantly used EPC tags in practice; 128-bit EPC tags have now started to appear in the marketplace with 256-bit EPC tags in the specification/prototype stage. Note that a 96-bit EPC is sufficient for most supply-chain operations (for reasons stated later). The structure of EPC as prescribed by the EPC-global specification primarily consists of four parts that corresponds to the preceding attributes:

- Header that denotes the EPC version used

- Manager Number that specifies the company name or the domain

- Object Class that represents the class type of the tagged object

- Serial Number, which as the name suggests, is the instance number of the tagged object

Figure 10-1 shows these fields of a 96-bit EPC.

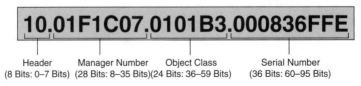

Figure 10-1 A 96-bit EPC.

An EPC can also incorporate an optional filter value based on which EPCs of tagged objects can be filtered in an efficient manner. Using 96 bits, you can generate a total of 79,228,162,514,264,337,593,543,950,336 (or about 80,000 trillion trillion) unique numbers! Another way of looking at a 96-bit EPC is that it can provide unique identifiers for 268 million companies with each company able to represent up to 16 million object classes and up to 68 billion unique serial numbers for each such object class.

Note that an EPC is strictly a unique identifier *and nothing else*. Therefore, *any* product-specific information has to reside separately in the enterprise back-end systems.

The following subsection discusses a very important concept called *EPC Class* tags.

10.4.1.1.1 EPC Class Tags

EPCglobal has defined the following four classes of EPC RFID tags to provide a range of capabilities at different price points. You should become intimately familiar with this classification because this is one of the core concepts that you might soon be using on a daily basis. Chapter 1, "Technology Overview," explains the terms used here:

- Class 0/Class 1
- Class 2
- Class 3
- Class 4

The following subsections discuss these classifications in detail.

10.4.1.1.1.1 EPC Class 0/Class 1

Both of these tag types are passive tags that can store either 64 bits or 96 bits of EPC data. A Class 0 tag data consists of a unique serial number that has already been written by the manufacturer before this tag is shipped to a customer. Class 0+ and Class 1 are WORM tags that allow data to be written by a customer at the point of use. Class 0 is defined for UHF (900 MHz) whereas Class 1 is defined for both UHF (860–930 MHz) and HF (13.56 MHz). All these tag types use backscatter technology for reader-to-tag communication. The tags are beam powered. These are the cheapest tag types available. Currently, Class 0 and Class 1 tags are not interoperable. (That is, a reader that can read a Class 0 tag might not be able to read a Class 1 tag and vice versa.)

A *UHF Generation 2* tag (often referred to as *EPC Gen 2* or simply *Gen 2* tag) is a new generation of EPC WORM tags based on the UHF Generation 2 Foundation Protocol that will replace the Class 0 and Class 1 tags. The specification was ratified as an EPC standard by EPCglobal on December 16, 2004. A Gen 2 tag is defined for UHF (860–930 MHz) and will consist of a 128-bit RW tag with 96 bits reserved for EPC data and 32 bits for error correction and a kill command. Gen 2 products are expected to appear on the market in the near future.

10.4.1.1.1.2 EPC Class 2

This is a passive RW tag that can store an EPC together with user data. The minimum user data capacity of such a tag is 224 bits. This tag uses backscatter technology for reader-to-tag communication. A Class 2 tag is beam powered. These are the next-cheapest tag types after Class 0/Class1. These tag types are still in the prototypical stage.

10.4.1.1.1.3   EPC Class 3

This is a RW active tag that has a large user data capacity that is not specified at this time. A Class 3 EPC tag supports on-board processing and I/O capability. This tag uses backscatter technology for reader-to-tag communication and is transmitter powered. These are the next-cheapest tag types after Class 2.

Class 3 tags are yet to be produced even for prototypical use.

10.4.1.1.1.4   EPC Class 4

This is a RW active tag with a large user data capacity that is yet to be specified. It supports on-board processing and I/O capability. This tag uses transmitter technology for reader-to-tag communication and is battery powered. The minimum read range is 300 feet (about 91 meters). These are the most expensive tag types.

Class 4 tags are yet to be produced even for prototypical use.

10.4.1.2   Data-Collection Hardware

The EPCglobal has already released specifications for EPC tags and interface protocols based on which readers and tags can be interoperable from different vendors. For example, a Class 1 EPC tag from one vendor could be read by a Class 1 EPC compatible reader from another vendor. This open nature provides great flexibility and promotes competition among different vendors to bring out superior products at a cheaper cost.

10.4.1.3   Discovery Services

This suite of services mediates and provides the access to EPC data. *Object Naming Service* (ONS) is a component of these services, and it is described in the following section.

10.4.1.3.1   Object Naming Service (ONS)

The ONS is a public service that can be used to find related EPCIS servers from where data about a product can be extracted. It provides a mapping mechanism between an EPC and the set of EPCIS instances that contain information about this EPC. (Thus, in essence, ONS is very similar to the DNS service that is used to look up the associated hosts for a particular Internet address.) The ONS service has to perform in real time so that it can quickly handle a very large number of requests (for example, in the trillions per day) reliably. In summary, the ONS is a service for extremely fast and reliable global database lookup.

10.4.1.4   EPCglobal Middleware

A tag can be read multiple times by the same or different readers at different points in the supply chain. Each such read generates tag data on the reader side and hence on the EPCglobal Network. As a result, a tremendous amount of data is generated on the EPCglobal Network as a result of reading the tags. A substantial portion of this data can be compressed because it might just consist of duplicate reads, reads that can be combined with other reads, and reads that are not significant in terms of business logic, and so on. If this data were stored and transported as is, most storage systems and networks would collapse. To handle this data efficiently, it needs to be sorted, filtered, and processed so that it can be managed in real time. This is the functionality of EPCglobal middleware. In addition to the previously described tasks, it is also responsible for movement of relevant information through the network to EPCIS or other business back-end systems of an enterprise. As a result, the data volume is reduced and data is transmitted selectively in the network, making the use of such data efficient and useful.

10.4.1.5 EPC Information Services (EPCIS)

These are gateways hosted by secure servers that contain information about items with EPC numbers in an EPCglobal Network. An EPCIS associates EPC data with business events and information. This is useful for automatically triggering execution of event-based logic. Several EPCIS instances can store the information about a single EPC number. Thus, information for a particular EPC is distributed in nature. Therefore, to assemble particular information about an EPC, data from several EPCIS instances might need to be extracted and merged. An EPCIS can act as a façade to a collection of business back-end systems, such as *warehouse management systems* (WMS), *enterprise resource planning* (ERP), and homegrown systems. EPCIS instances of a business should be shared among trading partners, suppliers, and clients to increase collaboration with these parties. Information from EPCIS is extracted in the form of *Physical Markup Language* (PML), which is described next.

10.4.1.5.1 Physical Markup Language (PML)

This is an open XML schema for representing product information as well as communication. Currently, PML can be divided into the following two parts:

- **Core PML.** The core components of EPCglobal Network use this open XML scheme to communicate with each other. This PML type already has been specified.

- **Extended PML.** The EPCglobal specification uses this open XML schema to represent the physical characteristics of products. This PML type has not been completely specified yet. Examples of this XML are item expiration date, location history, recycling information, composition information, manufacturing date, and so forth.

The next section discusses how these components work together to form the EPCglobal Network.

10.4.1.6 Tying It All Together

The EPC data on the tags is read by the readers. This data is then passed to the middleware for proper management via wired or wireless network. The Discovery Services provide the location information of the EPCIS instance to the middleware. The middleware adds location and event information to the processed data and moves it to the appropriate EPCIS instance for storage and action. Figure 10-2 shows this process.

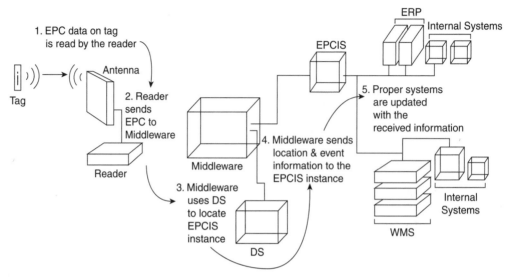

Figure 10-2   An EPCglobal Network schematic.

10.5   U.S. Department of Defense (DoD)

The DoD officially released its RFID policy on July 30, 2004. DoD contracts issued as of October 1, 2004, for material delivery on or after January 1, 2005, must use RFID tags. Starting January 1, 2007, all commodities and commodity pallets shipped to any DoD facility must have RFID tags. You can download the entire policy and related materials from www.dodrfid.org. Some of the policy highlights are as follows:

- Passive tags should be attached to pallets and cases. This also applies to individual high-value items that have a *unique identification* (UID) *code*.
- A supplier can use either the EPC or UID data format to encode item identity.
- Use of passive UHF tags operating between 860–960 MHz with a minimum read distance of 9 feet (about 2.7 meters). EPC Class 0 (64 and 96 bits) and EPC Class 1 (64 and 96 bits) are acceptable. These tags will be phased out after UHF Gen 2 tags and readers are available.
- All 20- and 40-foot containers shipped outside the United States should have an active tag containing the content list data written at the point of origin. This applies to both airborne and seagoing cargo containers.

  The DoD is already convinced of the potential benefits of RFID through multiple pilots run over a period of several months. It is expected that RFID will provide better inventory management and control. This can also translate into better released support for the troops in the battlefield.

10.6   ISO (International Organization for Standardization)

ISO is a network of the national standards institutes of 146 countries, on the basis of 1 member per country, with a Central Secretariat in Geneva, Switzerland, that coordinates the system. ISO is a nongovernmental organization.

ISO has the following TCs (*Technical Committees*) and JTCs (*Joint Technical Councils*) that are involved in formulating RFID-related standards:

- ISO JTC1 SC31
- ISO JTC1 SC17
- ISO TC 104 / SC 4
- ISO TC 23 / SC 19
- ISO TC 204
- ISO TC 122

The preceding list is not exhaustive. The following ISO standards are related to RFID technology and its use in real-world applications:

- **ISO 6346.** Freight containers—Coding, identification, and marking.
- **ISO 7810.** Identification cards—physical characteristics. Provides performance criteria, requirements for international exchanges, and minimum man-machine features criteria.
- **ISO 7816.** Identification cards—integrated circuit(s) cards with contacts. This is currently composed of the following 12 parts:

  Part 1. Physical characteristics
  Part 2. Dimensions and location of the contacts
  Part 3. Electronic signals and transmission protocols
  Part 4. Interindustry commands for interchange
  Part 5. Numbering system and registration procedure for application identifiers
  Part 6. Interindustry data elements for interchange
  Part 7. Interindustry commands for Structured Card Query Language (SCQL)
  Part 8. Commands for security operations
  Part 9. Commands for card management
  Part 10. Electronic signals and answer to reset for synchronous cards
  Part 11. Personal verification through biometric methods
  Part 12. Cryptographic information application

- **ISO 9798.** Information technology—Security techniques—Entity authentication. This is currently composed of the following five parts:

  Part 1. General
  Part 2. Mechanisms using symmetric encipherment algorithms
  Part 3. Mechanisms using digital signature techniques

Part 4. Mechanisms using a cryptographic check function

Part 5. Mechanisms using zero knowledge techniques

- **ISO 9897.** Freight containers—Container equipment data exchange (CEDEX)—General communication codes.

- **ISO 10373.** Identification cards—test methods. This is currently composed of the following six parts:

  Part 1. General characteristics tests

  Part 2. Cards with magnetic stripes

  Part 3. Integrated circuit(s) cards with contacts and related interface devices

  Part 5. Optical memory cards

  Part 6. Proximity cards

  Part 7. Vicinity cards

- **ISO 10374.** Freight Containers—Automatic Identification.

- **ISO 10536.** Identification cards—Contactless integrated circuit(s) cards—Close-coupled cards. This currently consists of the following three parts:

  Part 1. Physical characteristics

  Part 2. Dimensions and location of coupling areas

  Part 3. Electronic signals and reset procedures

- **ISO 11784.** Radio frequency identification of animals—Code structure. This specifies the identification code structure for identifying an animal using RFID. However, it does not specify any transmission protocol characteristics between an RFID tag and reader. This is provided by the next standard.

- **ISO 11785.** Radio frequency identification of animals—Technical concept. This specifies the transmission procedure between an RFID tag and reader.

- **ISO 14223.** Radio frequency identification of animals—Advanced transponders. This consists of the following part:

  Part 1. Air interface. This specifies the air interface between the RFID reader and the advanced transponder (tag) that is fully compatible with ISO 11784 and 11785.

- **ISO 14443.** Identification cards—Contactless integrated circuit(s) cards—Proximity cards. This is currently composed of the following four parts:

  Part 1. Physical characteristics

  Part 2. Radio frequency power and signal interface

  Part 3. Initialization and anticollision

  Part 4. Transmission protocol

- **ISO 14816.** Road Traffic and Transport Telematics—Automatic Vehicle and Equipment Identification—Numbering and Data Structures.

- **ISO 15434.** Information technology—Transfer syntax for high-capacity ADC media.

- **ISO 15459.** Information technology—Unique identification of transport units. This is composed of the following two parts:

  15459 Part 1. Technical standard
  15459 Part 2. Registration procedures

- **ISO 15961.** Information technology—Radio frequency identification (RFID) for item management—Data protocol: application interface.

- **ISO 15962.** Information technology—Radio frequency identification (RFID) for item management—Data protocol: data encoding rules and logical memory functions.

- **ISO 15963.** Information technology—Radio frequency identification for item management—Unique identification for RF tags. This describes numbering schemes for uniquely identifying RFID tags.

- **ISO 17358.** Supply chain application for RFID—Application requirements. This is currently under development.

- **ISO 17363.** Supply chain application for RFID—Freight containers. This is currently under development.

- **ISO 17364.** Supply chain application for RFID—Transport units. This is currently under development.

- **ISO 17365.** Supply chain application for RFID—Returnable transport items. This is currently under development.

- **ISO 17366.** Supply chain application for RFID—Product packaging. This is currently under development.

- **ISO 17367.** Supply chain application for RFID—Product tagging. This is currently under development.

- **ISO 18000.** Information technology—RFID for item management. This is currently composed of six parts:

  Part 1. Reference architecture and definition of parameters to be standardized
  Part 2. Parameters for air interface communications below 135 kHz
  Part 3. Parameters for air interface communications at 13.56 MHz
  Part 4. Parameters for air interface communications at 2.45 GHz
  Part 6. Parameters for air interface communications at 860 to 930 MHz
  Part 7. Parameters for active air interface communications at 433 MHz

- **ISO 18001.** Information technology—Radio frequency identification for item management—Application requirements profiles.

- **ISO 18046.** RFID Tag and Interrogator Performance Test Methods. This is yet to be published.

- **ISO 18047.** Information technology—Radio frequency identification device conformance test methods. This is split to mirror the ISO 18000 standard. This is currently composed of the following three parts:

  Part 3. Test methods for air interface communications at 13.56 MHz
  Part 4. Test methods for air interface communications at 2.45 GHz
  Part 7. Test methods for air interface communications at 433 MHz

- **ISO 18185.** Freight containers—Radio frequency communication protocol for electronic seal. This is under development. This is composed of the following seven parts:

  Part 1. Communication protocol
  Part 2. Application requirements
  Part 3. Environmental characteristics
  Part 4. Data protection
  Part 5. Sensor interface
  Part 6. Message sets for transfer between seal reader and host computer
  Part 7. Physical layer

- **ISO 19762.** Information technology AIDC techniques—Harmonised Vocabulary. This is under development. This consists of the following three parts:

  Part 1. General terms relating to AIDC
  Part 2. Optically readable media (ORM)
  Part 3. Radio frequency identification (RFID)

- **ISO 23389.** Freight containers—read-write radio frequency identification (RFID).

- **ISO 24710.** Information technology AIDC techniques—RFID for Item Management—ISO 18000 Air Interface Communications—Elementary Tag license-plate functionality for ISO 18000 air interface definitions.

10.7   ETSI (European Telecommunications Standards Institute)

ETSI is an independent, nonprofit organization in Europe whose mission is to develop telecommunications standards for today and for the future.

The following ETSI standards are relevant to RFID:

- **ETSI TR 101 445 V1.1.1 (2002-04).** Electromagnetic Compatibility and Radio spectrum Matters (ERM); Short Range Devices (SRD) intended for operation in the 862 MHz to 870 MHz band; System Reference Document for radio frequency identification (RFID) equipment.

- **ETSI I-ETS 300 220 ed.1 (1993-10).** Radio Equipment and Systems (RES); Short Range Devices (SRD); Technical characteristics and test methods for radio equipment to be used in the 25 MHz to 1,000 MHz frequency range with power levels ranging up to 500 mW. This is currently composed of the following three parts:

Part 1. V1.3.1 (2000-09). Technical characteristics and test methods
Part 2. V1.2.1 (1997-11). Supplementary parameters not intended for regulatory purposes
Part 3. V1.1.1 (2000-09). Harmonised EN covering essential requirements under article 3.2 of the R&TTE Directive

- **ETSI EN 300 330 V1.2.2 (1999-05).** ElectroMagnetic Compatibility and Radio Spectrum Matters (ERM); Short Range Devices (SRD); Technical characteristics and test methods for radio equipment in the frequency range 9 kHz to 25 MHz and inductive loop systems in the frequency range 9 KHz to 30 MHz. This is currently composed of the following two parts:

Part 1. V1.4.1 (2004-11). Technical characteristics and test methods
Part 1. V1.2.1 (2004-11). Harmonised EN under article 3.2 of the R&TTE Directive

- **ETSI I-ETS 300 440/C1 ed.1 (1996-04).** Radio Equipment and Systems (RES); Short Range Devices (SRD); Technical characteristics and test methods for radio equipment to be used in the 1 GHz to 25 GHz frequency range. This is currently composed of the following two parts:

ETSI EN 300 440-1 V1.3.1 (2001-09). Technical characteristics and test methods
ETSI EN 300 440-2 V1.1.2 (2004-07). Harmonised EN under article 3.2 of the R&TTE Directive

- **ETSI EN 300 674 V1.1.1 (1999-02).** Electromagnetic Compatibility and Radio Spectrum Matters (ERM); Road Transport and Traffic Telematics (RTTT); Technical characteristics and test methods for Dedicated Short Range Communication (DSRC) transmission equipment (500 kbps / 250 kbps) operating in the 5,8 GHz Industrial, Scientific, and Medical (ISM) band. This is currently composed of the following two parts:

ETSI EN 300 674-1 V1.2.1 (2004-08). General characteristics and test methods for Road Side Units (RSU) and On-Board Units (OBU)
ETSI EN 300 674-2-1 V1.1.1 (2004-08). Harmonised EN under article 3.2 of the R&TTE Directive

- **ETSI ETS 300 683 ed.1 (1997-06).** Radio Equipment and Systems (RES); Electro-Magnetic Compatibility (EMC) standard for Short Range Devices (SRD) operating on frequencies between 9 KHz and 25 GHz.

- **ETSI EN 300 761 V1.1.1 (1998-01).** Electromagnetic Compatibility and Radio Spectrum Matters (ERM); Automatic Vehicle Identification (AVI) for railways. This is currently composed of the following two parts:

ETSI EN 300 761-1 V1.2.1 (2001-06). Technical characteristics and methods of measurement
ETSI EN 300 761-2 V1.1.1 (2001-06). Harmonised standard covering essential requirements under article 3.2 of the R&TTE Directive

- **ETSI EN 301 489.** Electromagnetic Compatibility and Radio Spectrum Matters (ERM); ElectroMagnetic Compatibility (EMC) standard for radio equipment and services. This is currently composed of the following 30 parts:

ETSI EN 301 489-1 V1.5.1 (2004-11). Common technical requirements

ETSI EN 301 489-2 V1.3.1 (2002-08). Specific conditions for radio paging equipment

ETSI EN 301 489-3 V1.4.1 (2002-08). Specific conditions for Short Range Devices (SRD) operating on frequencies between 9 KHz and 40 GHz

ETSI EN 301 489-4 V1.3.1 (2002-08). Specific conditions for fixed radio links and ancillary equipment and services

ETSI EN 301 489-5 V1.3.1 (2002-08). Specific conditions for Private land Mobile Radio (PMR) and ancillary equipment (speech and nonspeech)

ETSI EN 301 489-6 V1.2.1 (2002-08). Specific conditions for Digital Enhanced Cordless Telecommunications (DECT) equipment

ETSI EN 301 489-7 V1.2.1 (2002-08). Specific conditions for mobile and portable radio and ancillary equipment of digital cellular radio telecommunications systems (GSM and DCS)

ETSI EN 301 489-8 V1.2.1 (2002-08). Specific conditions for GSM base stations

ETSI EN 301 489-9 V1.3.1 (2002-08). Specific conditions for wireless microphones, similar radio frequency (RF) audio link equipment, cordless audio and in-ear monitoring devices

ETSI EN 301 489-10 V1.3.1 (2002-08). Specific conditions for First (CT1 and CT1+) and Second Generation Cordless Telephone (CT2) equipment

ETSI EN 301 489-11 V1.2.1 (2002-11). Specific conditions for terrestrial sound broadcasting service transmitters

ETSI EN 301 489-12 V1.2.1 (2003-05). Specific conditions for Very Small Aperture Terminal, Satellite Interactive Earth Stations operated in the frequency ranges between 4 GHz and 30 GHz in the Fixed Satellite Service (FSS)

ETSI EN 301 489-13 V1.2.1 (2002-08). Specific conditions for Citizens' Band (CB) radio and ancillary equipment (speech and nonspeech)

ETSI EN 301 489-14 V1.2.1 (2003-05). Specific conditions for analog and digital terrestrial TV broadcasting service transmitters

ETSI EN 301 489-15 V1.2.1 (2002-08). Specific conditions for commercially available amateur radio equipment

ETSI EN 301 489-16 V1.2.1 (2002-08). Specific conditions for analog cellular radio communications equipment, mobile and portable

ETSI EN 301 489-17 V1.2.1 (2002-08). Specific conditions for 2,4 GHz wideband transmission systems and 5 GHz high-performance RLAN equipment

ETSI EN 301 489-18 V1.3.1 (2002-08). Specific conditions for Terrestrial Trunked Radio (TETRA) equipment

ETSI EN 301 489-19 V1.2.1 (2002-11). Specific conditions for Receive Only Mobile Earth Stations (ROMES) operating in the 1,5 GHz band providing data communications

ETSI EN 301 489-20 V1.2.1 (2002-11). Specific conditions for Mobile Earth Stations (MES) used in the Mobile Satellite Services (MSS)

ETSI EN 301 489-22 V1.1.1 (2003-11). Specific conditions for ground-based VHF aeronautical mobile and fixed radio equipment

ETSI EN 301 489-23 V1.2.1 (2002-11). Specific conditions for IMT-2000 CDMA Direct Spread (UTRA) Base Station (BS) radio, repeater and ancillary equipment

ETSI EN 301 489-24 V1.2.1 (2002-11). Specific conditions for IMT-2000 CDMA Direct Spread (UTRA) for Mobile and portable (UE) radio and ancillary equipment

ETSI EN 301 489-25 V2.3.2 (2004-07). Specific conditions for CDMA 1x Spread Spectrum Mobile Stations and ancillary equipment

ETSI EN 301 489-26 V2.3.2 (2004-07). Specific conditions for CDMA 1x Spread Spectrum Base Stations, repeaters and ancillary equipment

ETSI EN 301 489-27 V1.1.1 (2004-06). Specific conditions for Ultra Low Power Active Medical Implants (ULP-AMI) and related peripheral devices (ULP-AMI-P)

ETSI EN 301 489-28 V1.1.1 (2004-09). Specific conditions for wireless digital video links

ETSI EN 301 489-31 V1.1.1 (2004-11). Specific conditions for equipment in the 9 to 315 KHz band for Ultra Low Power Active Medical Implants (ULP-AMI) and related peripheral devices (ULP-AMI-P)

ETSI EN 301 489-32 V1.1.1 (2004-11). Specific conditions for Ground and Wall Probing Radar applications

ETSI EN 301 489-28 V1.1.1 (2004-09). Specific conditions for wireless digital video links

10.8   ERO (European Radiocommunications Office)

ERO supports ECC of the CEPT. ECC is the committee that brings together the radio- and telecommunications regulatory authorities of the 45 CEPT member countries.

The following list identifies ERO activities related to RFID:

- **ECC Report 001.** Compatibility between inductive LF and HF RFID transponder and other radio communications systems in the frequency ranges 135–148.5 kHz, 4.78–8.78 MHz, and 11.56–15.56 MHz.

- **ECC Report 007.** Compatibility between inductive LF RFID systems and radio communications systems in the frequency range 135–148.5 kHz.

- **ERC Report 074.** Compatibility between radio frequency identification devices (RFID) and the radioastronomy service at 13 MHz.

- **ERC/DEC(91)01.** ERC Decision of 12 March 2001 on harmonised frequencies, technical characteristics and exemption from individual licensing of Non-specific Short Range Devices operating in the frequency bands 6,765–6,795 kHz and 13.553–13.567 MHz.

- **ERC/DEC(91)02.** ERC Decision of 12 March 2001 on harmonised frequencies, technical characteristics and exemption from individual licensing of Non-specific Short Range Devices operating in the frequency band 26.957–27.283 MHz.

- **ERC/DEC(91)03.** ERC Decision of 12 March 2001 on harmonised frequencies, technical characteristics and exemption from individual licensing of Non-specific Short Range Devices operating in the frequency band 40.660–40.700 MHz.

- **ERC/DEC(91)04.** ERC Decision of 12 March 2001 on harmonised frequencies, technical characteristics and exemption from individual licensing of Non-specific Short Range Devices operating in the frequency bands 868.0–868.6 MHz, 868.7–869.2 MHz, 869.4–869.65 MHz, 869.7–870.0 MHz.

- **ERC/DEC(91)05.** ERC Decision of 12 March 2001 on harmonised frequencies, technical characteristics and exemption from individual licensing of Non-specific Short Range Devices operating in the frequency band 2,400–2,483.5 MHz.

- **ERC/DEC(91)08.** ERC Decision of 12 March 2001 on harmonised frequencies, technical characteristics and exemption from individual licensing of Short Range Devices used for Movement Detection and Alert operating in the frequency band 2,400–2,483.5 MHz.

- **ERC/DEC(91)13.** ERC Decision of 12 March 2001 on harmonised frequencies, technical characteristics and exemption from individual licensing of Short Range Devices used for inductive applications operating in the frequency bands 9–59.750 kHz, 59.750–60.250 kHz, 60.250–70 kHz, 70–119 kHz, 119–135 kHz.

- **ERC/DEC(91) 14.** ERC Decision of 12 March 2001 on harmonised frequencies, technical characteristics and exemption from individual licensing of Short Range Devices used for inductive applications operating in the frequency bands 6,765–6,795 kHz, 13.553–13.567 MHz.

- **ERC/DEC(91) 15.** ERC Decision of 12 March 2001 on harmonised frequencies, technical characteristics and exemption from individual licensing of Short Range Devices used for inductive applications operating in the frequency band 7,400–8,800 kHz.

- **ERC/DEC(91) 16.** ERC Decision on 12 March 2001 on harmonised frequencies, technical characteristics and exemption from individual licensing of Short Range Devices used for inductive applications operating in the frequency band 26.957–27.283 MHz.

- **ERC/DEC(92)02.** ERC Decision of 22 October 1992 on the frequency bands to be designated for the coordinated introduction of Road Transport Telematic Systems.

- **ERC/REC 70-03.** (Recommendation) Procedure for mutual recognition of type testing and type approval for radio equipment.

- **ERC/REC 70-03.** (Recommendation) Relating to the use of Short Range Devices (SRD).

10.9   The Open Services Gateway Initiative

The specification from this popular organization (consisting of about 60 member companies) is not RFID specific per se, but can be used for management of RFID edge systems and controllers (see Chapter 1). The following summarizes the main aspects of this organization and its service platform specification:[1]

> *The Open Services Gateway Initiative (OSGi) was founded in March 1999. Its mission is to create open specifications for the network delivery of managed services to local networks and devices. The OSGi service platform specification provides an open, common architecture for service providers, developers, software vendors, gateway operators and equipment vendors to develop, deploy and manage services in a coordinated fashion. It enables an entirely new category of smart devices due to its flexible and managed deployment of services. The primary targets for the OSGi specifications are set top boxes, service gateways, cable modems, consumer electronics, PCs, industrial computers, cars and more. These devices that implement the OSGi specifications will enable service providers like telcos, cable operators, utilities, and others to deliver differentiated and valuable services over their networks.*

Although this service is primarily developed for service gateways and so forth, it can be used with the same effectiveness for RFID edge devices.

10.10   Contact Information of Standards Bodies

The following list provides the contact information of the standards organizations:

- **ANSI (American National Standards Institute).** www.ansi.org. 25 West 43rd Street, 4th floor, New York, NY 10036, USA. Phone: 212-642-4900. Fax: 212-398-0023. E-mail: info@ansi.org.
- **Automotive Industry Action Group (AIAG).** www.aiag.org. 26200 Lahser Road, Suite 200, Southfield, MI 48034-7100. Phone: 248-358-3003. Fax: 248- 799-7995. E-mail: order_inquiry@aiag.org.
- **EAN.UCC (European Article Numbering Association International, Uniform Code Council).** Uniform Code Council, Inc. www.uc-council.org. Princeton Pike Corporate Center, 1009 Lenox Drive, Suite 202, Lawrenceville, NJ 08648. Phone: 609-620-0200. Fax: 609-620-1200. E-mail: info@uc-council.org.
- **EPCglobal.** www.epcglobalinc.com. Princeton Pike Corporate Center, 1009 Lenox Drive, Suite 202, Lawrenceville, NJ 08648. Phone: 609-620-4671. Fax: 609-620-0255. E-mail: EPCInfo@EPCglobalUS.org.

---

[1]   OSGi Service Platform. Release 3. March 2003, pg. 3.

- **ISO (International Organization for Standardization).** www.iso.org. For sales enquiries, etc., access the web site. You can access the online standards catalog at www.iso.org/iso/en/CatalogueListPage.CatalogueList.

- **CEN (Comité Européen de Normalisation (European Committee for Standardization).** www.cenorm.be. 36 rue de Stassart, B - 1050 Brussels, Belgium. Phone: + 32 2 550 08 11. Fax: + 32 2 550 08 19. E-mail: infodesk@cenorm.be.

- **ETSI (European Telecommunications Standards Institute).** www.etsi.org. 650 route des Lucioles 06921 Sophia-Antipolis Cedex, France. Phone: +33 (0)4 92 94 42 00. Fax: +33 (0)4 93 65 47 16. E-mail: helpdesk@etsi.org.

- **ERO (European Radiocommunications Office).** www.ero.dk. Peblingehus, Nansens-gade 19, DK 1366   Copenhagen. Phone: +45 33896300. Fax: +45 33896330. E-mail: ero@ero.dk.

- **UPU (Universal Postal Union).** www.upu.int. Case postale 13, 3000 BERNE 15, SWITZERLAND. Phone: +41 31 350 31 11. Fax: +41 31 350 31 10. E-mail: info@upu.int.

- **ASTM (American Society for Testing and Materials).** www.astm.org 100 Barr Harbor Drive, West Conshohocken, PA 19428-2959. Phone: (610) 832-9585. Fax: (610) 832-9555. E-mail: service@astm.org.

- **OSGi Alliance.** www.osgi.org. Bishop Ranch 6, 2400 Camino Ramon, Suite 375, San Ramon, CA 94583 USA. Phone: (925) 275-6625. Fax: (925) 886-3696. E-mail: help@osgi.org.

CHAPTER 11

Closing Thoughts

RFID is not a silver bullet, but it is here to stay (and its use will continue to grow). Therefore, even if a business has not yet attempted to evaluate the benefits of RFID (in its business context), it cannot ignore the technology all together. A business might gain early cost savings and advantages over its competitors if it seizes the opportunity to perform this evaluation now. Remember, however, that in spite of the current excitement regarding the technology, RFID might not be suitable for everyone, and it is not an endeavor to take lightly. This book would not be complete without a discussion of the current hurdles often cited as the principal reasons for not using RFID. The next section addresses these hurdles. After that, a few brief comments and observations about RFID technology close this chapter and this book.

11.1 Current Hurdles to Using RFID

The following are some of the main hurdles cited for not using RFID in a business:

- Too expensive
- Too complex/does not work
- Existing solutions suffice
- Does not offer any business advantages
- Does not apply

The following subsections take a closer look at these reasons and some ways to bypass these roadblocks.

11.1.1   Too Expensive

This is probably the most common factor cited for not using RFID. Although RFID tags and readers are not some of the cheapest items available today, the prices are falling rapidly, and the technology is improving at an impressive rate. Therefore, if RFID is dismissed today solely based on this factor, the business is strongly advised to revisit its business cases at least once in the next two years. In addition, in this context, it is important to determine the tipping point, or price threshold of the system, at which the business can afford an RFID solution. Figure 11-1 shows an example scenario.

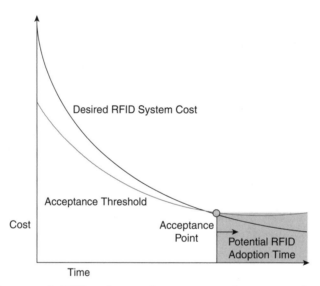

Figure 11-1   An example RFID system cost-versus-acceptance scenario.

Here, the tipping point, or the price threshold, is at the intersection point marked Acceptance Point of the two curves representing the cost of the RFID system needed versus its acceptable cost. The evaluation and adoption of RFID can begin after this point in time (that is, the right side of the line passing through the acceptance point as shown shaded). This determination is especially necessary in the situation where the business benefits of RFID are already apparent but the cost presents a hurdle. It might be possible to negotiate a price point that is either at or below the threshold with a vendor partner based on certain factors (for example, a large volume order). This option can thus alleviate the price issue without any need for waiting for prices to come down. Whereas every business might want a sub-1¢ tag, this desire raises one major issue: It might take a long time for prices to drop that low. For example, it might happen in another 10, 15, or 20 years, assuming it happens at all. Therefore, if a business waits for a sub-1¢ tag when its competitor has figured out a more realistic cost threshold, the former will be at a serious competitive disadvantage if the latter can successfully exploit the technology benefits when its cost threshold is met.

If the current price of RFID hardware is the main concern, a business should determine a realistic price threshold at which it can afford an RFID solution. It should then revisit its business cases after this threshold has been met.

11.1.2   Too Complex/Does Not Work

This conclusion sometimes emerges after the usual business justification and pilot implementation have been completed to evaluate the technology. Although the business reasons to use RFID might be compelling, the actual implementation of a test pilot might reveal several issues involving complexity and cost. Most of these issues can be traced back to two things: state of immaturity of the technology and the current cost of RFID hardware. For example, perhaps no suitable tags are currently available on the market with which to tag the desired items satisfactorily. Building a custom tag is an expensive proposition that can cost in the range of hundreds of thousands of dollars. However, vendors can customize existing tags (for example, by modifying the tag antenna of one or more of its existing tags) to eliminate any need for custom tags. In addition, an RFID system involves several variables that require time to experiment and optimize (see Chapter 9, "Designing and Implementing an RFID Solution"). This can frustrate someone who is looking for a plug-and-play, quick solution. Also, it is a good idea to seek out the advice of experts who have considerable experience in implementing real-world RFID solutions.

Patience is a much-needed virtue when it comes to implementing an RFID solution, especially in the beginning stages when it perhaps hits a roadblock. In such situations, you should always seek external help from people who have real-world RFID implementation experience.

11.1.3   Existing Solutions Suffice

That existing solutions exist is probably the third most common reason cited for not using RFID. However, a business justification for RFID is at least warranted before reaching such a conclusion. Otherwise, without objective data to rule out RFID, the business might just be exhibiting an "if it ain't broke, don't fix it" mentality, which might prove detrimental to the business (especially if competitors, customers, and business partners are eager to evaluate and use the technology to their advantage). In fact, such eager colleagues might change the rules drastically. For example, if a major customer demands the business to be RFID-enabled within a certain amount of time, noncompliance risks losing this client and hurting the bottom line (not to mention losing market share to competitors). In this case, the business has little choice but to move forward with some form of RFID adoption activity (perhaps even in spite of the fact that the business justification for using RFID for the business itself might not be so compelling).

Even if the business has not received an RFID-compliance mandate from its major customers, business partners, or suppliers, it should at least perform a business justification before deciding against using RFID. Even a small improvement that results from use of the technology (for example, a 5 percent reduction in inventory costs) can lead to substantial accrued benefit in the medium to long term (see Figure 11-2).

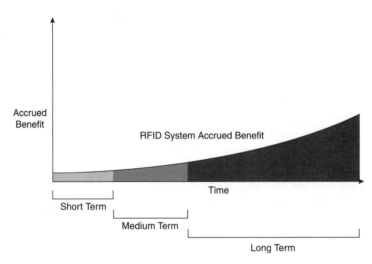

Figure 11-2   Potential accrued benefits of an example RFID system.

In addition, the use of RFID might itself lead to the discovery of unexpected areas of benefit not previously apparent.

11.1.4   Does Not Offer Any Business Advantages

This reason is also mentioned for not using RFID, and might actually mean two different things: First, RFID does not offer compelling business advantages based on its *perceived* potential; second, RFID cannot offer the needed benefits after the due diligence of business justification and evaluation pilots have been completed. The latter is the correct albeit the "hard" approach to objectively ruling out RFID. In contrast, the former might result from lazy "thinking" (or not thinking), basing an opinion on unrealistic figures, "anecdotal" evidence, "similar" case studies, and sweeping opinions in the media while bypassing the actual usefulness and impact of RFID on the business itself. The conclusions reached by such an analysis should be taken with several grains of salt, irrespective of its recommendation about the use of RFID. Why? Because such an analysis will most probably be too unrealistic or too pessimistic about the benefits that can be achieved from the use of the technology.

An objective analysis of the benefits of RFID and some level of actual evaluation are needed in the actual context of the business before accepting or discounting the use of the technology.

11.1.5   Does Not Apply

Although a business might rely heavily on data collection (and, remember, RFID is a data-collection technology) and the use of that data, that business might have concluded (at this point, anyway) that RFID does not apply to its areas of operation. Remember, however, that the potential uses of RFID technology are expanding and restricted neither to the supply chain nor to the examples provided in Chapter 4, "Application Areas." Therefore, although this business might be in that rare situation in which RFID, in fact, "does not apply," that situation might turn out to be an unexplored area of RFID application. Just because RFID is not applied to a certain area today does not mean it will not be applied tomorrow. Creative thinking and technological advances are the two major drivers of innovative RFID application potential.

Instead of giving up in despair, a business can either generate its own ideas or partner with a research organization to develop ideas. The business can then share these ideas with the vendors to create prototypes. With these prototypes, the business can quickly validate the effectiveness of the technology in the areas of interest. A successful outcome of such an endeavor might give the business a solid advantage over its competitors.

11.2 Comments and Observations

The following are a collection of comments and observations, in no particular order, about the current state of RFID technology:

- High degree of confusion.
- Exclusive in the short run.
- Inclusive in the medium to long run.
- No one has all the expertise.
- Everyone is still learning.
- Getting involved early is the best idea.

The following subsections examine each of these thoughts.

11.2.1 High Degree of Confusion

RFID today is characterized by hype and unbounded expectations on the one hand and sweeping negative opinions on the other. This dichotomy is enough to confuse business decision makers, let alone common people. Examples of such unrealistic comments are, "RFID will replace bar codes soon," and "RFID can be used to track anything anywhere in the world." Examples of negative opinions are, "50 percent of all RFID projects fail," and "RFID can be made so small that the tags can be put on people to track them from satellites." This book has attempted to correct several of these half-truths and hearsays. Chapter 6, "RFID Versus Bar Code," answers the first myth, and the very beginning section of Chapter 1, "Technology Overview," provides examples of where RFID cannot be used. The statistics concerning RFID project failure are bogus, but you must exercise extreme caution when citing or refuting (or especially when depending on) information (results, statistics, anecdotal data, and so on) from real-world RFID projects. Such accurate information is currently few and far between; because the technology is still being developed, assessed, and adopted, it is not uncommon for parties to closely guard actual results. However, because the craving for such information is so great, "anecdotal" evidence and results from severely scoped evaluation efforts are used to sate this appetite. As such data passes from person to person, it has a tendency to change completely. For example, a study that concludes that "50 percent of four test trials to track metal parts inventory in real time using UHF frequency succeeded" might undergo metamorphosis and end up as the myth cited previously. Therefore, always ask the following questions about any RFID statistics you encounter:

- Who conducted the study?
- What was the scope?
- What were the constraints?
- Does it apply to me?
- If so, how?

In addition, you should seek out the original study or report and refuse to accept second-hand reporting. From the original study, you can draw your own conclusions.

Finally, confusion abounds regarding passive and active tags and their capabilities. A passive tag reflects the signal of a reader to transmit its data, whereas an active tag does not and can act as a transmitter. Because active tags must carry a battery and on-board electronics, these are much larger and bulkier than passive tags. However, a common misconception holds that passive tags can act as transmitters, too. Although this misconception might seem harmless, it raises baseless fears and suspicions in people's minds regarding RFID technology. For example, Danish pop artist Jakob S. Boeskov and industrial designer Kristian von Bengtson created a hoax about a fake Danish sniper rifle called ID Sniper that can be used to shoot and implant miniature RFID tags into people without their knowledge, who can then be tracked using GPS (see Figure 11-3).

Figure 11-3   Danish ID Sniper rifle.

Reprinted with permission from Jakob S. Boeskov and Kristian von Bengtson

A bogus Web site called Empire North was created to pose as the manufacturer of this fiendish weapon. As you can immediately understand, active tags (which are much larger than passive tags) would be necessary in such a scenario. Therefore, the impact on a person being so tagged would be about the same as, or worse than, being hit by a bullet. The impact could destroy the tag or immobilize (perhaps permanently) the person hit, thus rendering any tracking feature dubious in value.

Facts aside now, consider some of the hysteria this prank created. A message on the popular Web site Slashdot drove more than a million and half people to the Empire North site. *Computerworld* ran an article despising this horrendous weapon. I even witnessed one of my friends searching eBay to buy this weapon for personal use! For Boeskov, this was a little more unnerving. Police from China and arms dealers from several countries around the world made approaches to buy this weapon! The sad fact that so many people believed in the sniper gun shows the dire necessity of educating the public about RFID.

11.2.2 Exclusive in the Short Run

Slap-and-ship type of implementations might turn out to be the most popular RFID solutions rolled out in the short run for various reasons, including the following:

- High cost
- Immature technology
- Implementation issues
- Incomplete understanding of process issues

An analysis of these factors follows. You can see that these factors substantially overlap those mentioned previously for not using RFID in a business today.

11.2.2.1 High Cost

Even with falling prices, RFID hardware remains expensive for most businesses, even for a slap-and-ship type application. Many small to medium-size businesses might not be able to justify the cost of such a system in the short run when the business justification for RFID might not be enough to offset this cost. In these cases, the only driver will be meeting some kind of an RFID mandate (example, for a customer), and thus probably spending minimum resources and effort (which typically means a slap-and-ship type deployment).

11.2.2.2 Immature Technology

That RFID is an immature technology is not in debate, and the following list identifies some issues that illustrate this point:

- **RFID hardware is in a flux.** The tag, antenna, and reader technology is improving at a rapid rate, which means that a business will have to upgrade hardware at one point or another. Such updates increase the cost of ownership of an RFID solution.

- **Hardware quality issues.** Defective tag rates are as high as 20 percent or more. In addition, not all tags perform equally well. Therefore, quality issues increase the hardware cost and can inhibit tag creation and readability, resulting in higher operating costs.

- **Standards are numerous and evolving.** Several RFID standards (data, tag, protocol, and middleware, for instance) exist, and several of these are still evolving, which might make selecting a standard or a set of standards (to implement and deploy an RFID solution) a difficult choice.

A business might decide to choose the path of least resistance and risk by implementing an RFID solution that is optimized in size and scope. The solution might also be isolated from other business processes as much as possible. Such a scenario seeks to minimize any impact on existing processes in case the solution malfunctions or needs to be replaced later. These factors point to a slap-and-ship type solution.

11.2.2.3   Implementation Issues

Significant implementation issues include the following:

- **Complexity.** An RFID solution implementation and deployment is anything but simple. Even with a slap-and-ship type application, the issues are several and involved.

- **Damaged hardware.** This is a real-world issue that is not going away any time soon. Reader antennas are probably the most fragile piece of RFID hardware, followed by readers, and then tags. In operating conditions where there is a high degree of rough handling, tags might be the most fragile piece. In these situations, tags can be ruggedized to lower the rate of damage (while increasing the hardware cost). In case of hardware failure, the defective hardware needs to be located and fixed or replaced. The system might need to be restarted. Business operations can be jeopardized as a result. A damaged tag presents a special problem because a reader cannot read the tagged item. If this item is part of a customer order, the supplier might not get paid for this item if the tag gets damaged before it arrives at the customer site.

- **Invalid tag reads.** A reader located close to a read zone might inadvertently read the "wrong" tag, resulting in a spurious read. This problem holds especially true for handheld readers that can read other tags in its read zone apart from the one it is supposed to read. In addition, a tag might not be read at all (see Chapter 9).

- **Interference from other radio equipment and RFID systems.** Operating conditions, often difficult to control, have an impact on RFID systems.

Rarely does one of these issues rise to the level of showstopper. However, time and cost are almost always involved to resolve the issues satisfactorily. A scoped-out RFID solution presents the least amount of challenge as far as the preceding variables are concerned. A slap-and-ship solution represents such a type of solution.

11.2.2.4 Incomplete Understanding of Process Issues

Process issues range from the simple to the extremely involved as the following list demonstrates:

- Steps need to be done before and after a downtime
- Process for fixing data errors such as an incorrect EPC in the system
- Process of replacing a damaged tag after it has been applied to an item
- Synchronizing processes for handling RFID and non-RFID items
- Process for exchanging data externally (for example, direct exchange or through a third party)

Unless a clear picture of these issues and their resolutions emerge, a business might not want to venture too far with an integrated RFID solution. In such cases, an isolated and limited-complexity solution might prove preferable. A slap-and-ship solution offers just that.

11.2.3 Inclusive in the Medium to Long Run

Although a business might prefer to deploy an isolated and exclusive solution to quickly meet the RFID mandates of its customers and reduce risk in the short term, the real benefit of RFID can be realized only when it is made *inclusive*. What does inclusive mean in this context? It means inclusive from all the following aspects:

- Systems
- Process
- Partners
- People

The following subsections explain these perspectives.

11.2.3.1 Systems Inclusive

Existing business systems should be integrated with the RFID system both for data and business transactions. In other words, RFID systems can be looked upon as an enterprise-wide infrastructure component. Existing systems should exploit the capabilities of this component to streamline, optimize, and interface with each other.

11.2.3.2 Process Inclusive

Although in the short run, the degree of integration of an RFID system with the business processes might be low, that integration stands a good chance of increasing when the different process issues are understood and resolved. The increased integration also represents how a business can realize the wider benefits of an RFID system. In addition, the business might uncover ways to streamline the processes and thus gain business efficiency and cost savings.

11.2.3.3   Partners Inclusive

To realize the full potential of an RFID system, a business must extend that system out to partners, customers, and suppliers. Therefore, the ultimate goal of an RFID system should be to enable collaboration between the business and its partners. Benefits derived from RFID such as anti-counterfeiting, inventory reduction, and reverse logistics (involving return of merchandise) can best be realized through collaboration among businesses. Collaboration, however, is time-consuming and involved. After all, different participating businesses might have different business and competitive interests and might want to focus on different parts of the value proposition. However, RFID implementation is not a mandatory prerequisite for collaboration to happen. You can use data from existing bar codes to build and validate such an effort. Standardized processes among businesses (for example, to automate shipping and receiving using *Advanced Ship Notice* [ASN]) are an important step. After some form of collaboration is in place, using RFID can make its potential benefits clear to the participating entities, which in turn can accelerate the degree of collaboration.

11.2.3.4   People Inclusive

Deploying an RFID solution involves cultural *change* in the business, so the deployment of such must take into account the human factor. This factor should be acknowledged as early as possible so that it does not create a cultural *clash*. The operations personnel might have baseless fears and anxiety about the technology, thinking of it as a threat to their privacy and job security. In addition, the personnel might have to modify their existing operating style to mesh with the RFID system. Therefore, training and education about the technology must be planned as a part of an RFID deployment strategy.

11.2.4   No One Has All the Expertise

RFID technology is segmented into many specialties, including the following:

- Tag microchip
- Tag antenna
- Tag
- Reader
- Reader antenna
- Edge system
- Middleware
- Applications specific to a business
- System integration
- Business consulting

No company in the RFID business today has all the answers (the expertise listed here). A business should pick the vendors that offer the best tradeoff to implement its RFID solution. In essence, the business has to play the role of an integrator. If the business is reluctant to take on this responsibility, the other option is to work with an experienced integrator and use it at as a single point of control. The business can then "shadow" the integrator team members to gain valuable insights and knowledge that will help it implement its next RFID application using in-house skills.

11.2.5   Everyone Is Still Learning

Besides the segmentation of RFID technology vendors, each vendor is in a continuous learning cycle to understand the technology, apply it, and then analyze the application results to further understand the nature of the technology to feed into the next iteration cycle. Therefore, the technology is improving at an impressive rate, but the RFID hardware is becoming outdated at a rapid rate, too. Besides the technology vendors, the early adopters are also in the learning mode. Although these businesses have already gained a considerable amount of RFID expertise through their evaluation pilots and small-scale deployments, they still have a long way to go to claim complete mastery of the technology.

11.2.6   Getting Involved Early Is the Best Idea

Currently, the early adopters of RFID technology enjoy a level of expertise unmatched by most businesses. After reading this book, you can understand that this technology takes time to master and that no one has complete mastery of the technology yet. These facts bolster the argument that businesses should initiate an RFID program as soon as possible. While the field of play is still relatively level, those that do initiate an RFID program will gain an advantage in understanding and using the technology. Businesses can take the initial steps (at least to some degree) of business justification and technology evaluation today. Even if an RFID deployment is not planned in the short term, it is important to have some degree of in-house knowledge and expertise in the technology. Such knowledge and expertise will come in handy when the business is ready to use RFID technology.

11.3   Conclusion

This book presented the facts about RFID technology in an unbiased manner from both theoretical and practical perspectives. The underlying goal was to provide you with a solid understanding, grounded in reality, of the different aspects of the technology. After reading this book, you should be able to comprehend and evaluate the merits of RFID without getting lost in the flux of opinions and media reports about this technology.

At this point, you have now learned a lot about crafting real-world RFID systems. You can use this knowledge to guide the implementation of your own RFID solutions. RFID is an important technology. A business cannot just ignore the vast potential of RFID and its impact. Sooner or later, every business (most likely) will have to deal with the technology in one way or another.

The business might first encounter RFID via its clients, suppliers, business partners, or competitors. Will the business be ready when that happens? What is the cost of not being ready? Can the business afford this cost? A business must answer these fundamental questions, among others, now that the technology is poised to take off. The answers might separate winners from losers. Although RFID is not perfect today, just remember that bar code technology was declared a failure in 1970s. Who knows what difference another 30 years will make for RFID! RFID is a journey forward to transform the world in ways never thought possible. Let this book serve as your trustworthy guide as you embark on this journey.

RFID Vendors, News Sources, and Conferences

This appendix lists resources related to product information, current events, and trends concerning the RFID industry. Because of the fast pace at which this industry is progressing, these lists of vendors, news sources, and conferences are by no means exhaustive.

> **NOTE**
>
> The vendors, news sources, and conferences are presented in alphabetical order and are provided for informational purposes only. A listed entity name does not represent implied or expressed endorsement of any kind whatsoever.

A.1   RFID Vendors

You might find the list of vendors provided here useful in various ways (for example, to make a first pass at selecting a set of vendors for an RFID project, to explore vendor product lines to find out about RFID system component current capabilities, to learn from vendor experiences, and to check out similar project references). This list is not exhaustive, and the listed vendor capabilities might have changed since the writing of this book. Therefore, to get a vendor's latest information, you should directly contact the vendor.

Figure A-1 provides a list of RFID vendors. The list columns in this figure are as follows:

- **Vendor.** Specifies a vendor's name and its Web address.
- **Tag.** Denotes whether the vendor designs or manufactures a tag microchip (M), tag/smart label (T), or tag antenna (A); provides other related services such as manufacturing an inlet or converting an inlet into a usable form (for example, smart labels, rolls, and

ruggedized tags) (O); and the tag type, either passive (Ps), active (Ac), or semi-active (Sa). Only the original vendors and their respective products are listed; resellers and partners are not included in this list.

- **Reader.** Specifies whether the vendor manufactures readers and printers of some type— for example, printer (P), stationary (S), handheld (H), serial [RS232/RS485] (R), wired network (N), or wireless (W).

- **Antenna.** Identifies whether the vendor designs or manufactures reader antennas.

- **Frequency.** Specifies the products' supported frequency types— LF (L), HF (H), UHF (U), and microwave (M).

- **Other Hardware/Software (HW/SW).** Indicates whether the vendor designs and implements RFID network, edge devices, or edge software (E); server, portal, tunnel, I/O controller, tag applicator/labeling system/RFID custom hardware, such as mounting hardware (H); middleware (M); or other software types (for example, application specific, tool sets, and business integration) (S).

- **Services.** Identifies whether the vendor provides services such as business consulting (B), RFID hardware consulting (H) (for example, site survey, installation, laboratory testing for tagging, label printing, and packaging services), integration (I), and training in the RFID technology (T).

Vendor	Tag M	T	A	O	Type Ps	Ac	Sa	Reader P	S	H	R	N	W	Antenna	Freq L	H	U	M	Other E	H	M	S	Serv B	H	I	T
Accenture www.accenture.com																							•	•	•	•
Accu-Sort Systems, Inc. www.accusort.com	•		•					•	•	•	•			•									•	•	•	•
Acsis, Inc. www.acsisinc.com			•																			•			•	•
ADT Security Services www.adt.com	•	•	•					•	•	•	•			•	•				•			•	•	•	•	•
Alien Technology Corporation www.alientechnology.com	•	•							•	•	•			•			•									
Automated Assembly Corporation www.autoassembly.com			•				•										•									
Avery Dennison RFID www.rfid.averydennison.com	•	•	•						•	•	•															
AWID www.awid.com	•	•	•					•	•	•	•	•				•	•									
Blue Vector Systems www.bluevectorsystems.com																										
Capgemini www.capgemini.com																										
Catalyst International www.catalystinternational.com																					•					
Checkpoint Systems, Inc. www.checkpointsystems.com	•	•	•					•	•	•	•				•	•					•					
Cisco Systems, Inc. www.cisco.com																						•				
CODEplus Software Engineering www.code-plus.com																					•	•				
ConnecTerra, Inc. www.connecterra.com																			•		•	•	•			
Cougaar Software, Inc. www.cougaarsoftware.com																					•					
The Danby Group www.danbygroup.com																										•
Datamax Corporation www.datamaxcorp.com	•	•	•		•			•												•						
Defense Systems, Inc. www.defensesys.com	•	•	•																							
A WFI Company																			•							•
Deloitte Consulting www.deloitte.com																										
ecVision, Inc. www.ecvision.com																										
Ekahau, Inc. www.ekahau.com															•					•	•		•	•	•	•
Escort Memory Systems http://www.ems-rfid.com	•	•	•						•	•	•			•						•						
EXTOL International, Inc. www.extol.com																					•					
FEIG Electronic, GmbH www.feig.de	•		•					•	•	•	•			•	•	•					•					•

Vendor	Tag				Type			Reader						Antenna	Frequency				Other HW/SW				Services			
	M	T	A	O	Ps	Ac	Sa	P	S	H	R	N	W		L	H	U	M	E	H	M	S	B	H	I	T
Franwell, Inc. www.franwell.com																							•	•	•	•
GenuOne, Inc. www.genuone.com																						•	•	•	•	•
Global eXchange Services, Inc. www.gxs.com																						•	•			
GlobeRanger, Inc. www.globeranger.com																				•	•	•		•	•	•
Hewlett-Packard www.hp.com																			•	•	•	•		•	•	•
HighJump Software a 3M Company www.highjump.com																										
Hitachi, Ltd. www.hitachi.com	•	•						•		•				•				•			•	•	•	•	•	•
IconNicholson www.iconnicholson.com																							•	•	•	•
IDVelocity, LLC www.idvelocity.com										•								•			•	•	•	•	•	•
Impinj, Inc. www.impinj.com	•	•														•							•	•	•	•
Infosys Technologies Limited www.infosys.com			•																							
Intermec Technologies Corp. www.intermec.com	•	•		•			•	•	•	•	•	•	•			•	•	•	•		•	•	•	•	•	•
International Business Machines Corporation www.ibm.com/solutions/rfid																										
Irista, Inc. www.irista.com An HK Systems Company				•																						
The Kennedy Group www.kennedygrp.com																			•							
Loftware, Inc. www.loftware.com																										
Lowry Computer Products, Inc. www.lowrycomputer.com			•					•	•							•			•	•		•		•	•	•
LXE, Inc. www.lxe.com																										
Manhattan Associates www.manh.com																						•	•	•	•	•
MARC Global www.marcglobal.com																										
MARKEM Applied Intelligence Solutions www.markem.com				•																•						•
Maxell Corporation of America www.maxell.com	•	•		•			•	•	•	•		•						•								
MET Laboratories, Inc. www.metlabs.com																								•		
Nashua Corporation www.nashuasmartlabels.com																								•		
NCR Corporation www.ncr.com																	•									
NJM/CLI Packaging Systems International www.njmcli.com																				•			•	•		•

Vendor	Tag				Type			Reader						Antenna	Frequency				Other HW/SW				Services			
	M	T	A	O	Ps	Ac	Sa	P	S	H	R	N	W		L	H	U	M	E	H	M	S	B	H	I	T
Northrop Grumman Corporation www.it.northropgrumman.com/rfid																			•					•	•	•
OATSystems, Inc. www.oatsystems.com																					•	•	•	•	•	
ODIN technolgies, Inc. www.odintechnologies.com																					•	•	•	•	•	•
ORACLE Corporation www.oracle.com/solutions/rfid																				•	•	•		•	•	
PAXAR Americas, Inc. www.paxar.com		•	•	•	•			•											•							
Philips Semiconductors www.semiconductors.philips.com	•	•	•	•										•		•	•									
Power Paper www.powerpaper.com		•	•	•	•									•			•									
Precisia LLC A wholly owned subsidiary of Flint Ink www.precisia.net			•		•								•			•	•		•				•	•		
Printronix, Inc. www.printronix.com		•	•	•	•			•									•			•	•					
Provia Software www.provia.com																						•	•			
PSC, Inc. www.psc.com									•	•	•	•														
R4 Global Solutions www.r4gs.com		•	•	•	•	•			•		•	•					•									
RF Code, Inc. www.rfcode.com		•	•	•		•					•	•					•									
RedPrairie, Inc. www.redprairie.com																						•	•			
SAMSys Technologies, Inc. www.samsys.com											•	•			•	•	•									
SAP America, Inc. www.sap.com																					•	•	•	•		
SATO America, Inc. www.satoamerica.com		•		•	•			•			•	•					•			•						
Savi Technology, Inc. www.savi.com	•			•	•			•			•	•					•			•	•	•	•	•	•	•
Science Applications International Corporation www.saic.com																						•	•	•	•	•
SeeBeyond Technology Corporation www.seebeyond.com																						•	•			
Ship2save www.ship2save.com																	•						•		•	
SIRIT Technologies, Inc. www.sirit.com			•			•					•	•				•	•									
SOKYMAT SA www.sokymat.com		•	•	•	•		•							•	•	•	•									
St. Onge Company www.stonge.com																						•	•		•	
Sun Microsystems www.sun.com/rfid		•	•	•	•				•	•	•	•					•				•	•	•	•	•	•
Symbol Technologies, Inc. www.symbol.com		•		•					•	•	•	•					•		•		•	•	•	•	•	•

Vendor	Tag M	T	A	O	Type Ps	Ac	Sa	Reader P	S	H	R	N	W	Antenna	Freq L	H	U	M	Other HW/SW E	H	M	S	Services B	H	I	T
TagsWare, Inc. www.tagsware.com	•	•	•																							
TEC America, Inc. www.tecamerica.com	•	•																	•		•		•	•	•	
Texas Instruments RFid Systems www.ti-rfid.com	•	•	•	•			•	•	•	•	•	•		•	•		•			•						
TIBCO www.tibco.com																				•						
TrailBlazer Systems, Inc. www.trailblazersystems.com																			•	•						
TRAXUS Technologies, Inc. www.traxustechnologies.com																			•	•						
Ubisense www.ubisense.net	•	•	•	•				•	•	•	•	•	•				•		•		•	•	•	•	•	•
UNISYS Corporation www.unisys.com																	•				•					
UPM Rafsec www.rafsec.com	•			•	•														•		•		•			
Velosel Corporation www.velosel.com															•						•					
Venture Research, Inc. www.ventureresearch.com															•							•				
Verisign, Inc. www.verisign.com/epc										•												•	•	•	•	•
Web Methods, Inc. www.webmethods.com																					•				•	•
Weber Marking Systems, Inc. www.webermarking.com																			•			•	•			
WJ Communications, Inc. www.wj.com								•	•	•	•	•	•		•	•	•	•	•		•					
X-ident USA LLC www.x-identusa.net																						•				
Xterprise, Inc. www.xterprise.com																						•				
Zebra, Inc. www.zebra.com								•							•				•		•		•	•	•	•

Figure A-1   A list of RFID vendors.

A.2   RFID News Sources

You can find articles about RFID in almost any print and electronic news source today. However, among these sources, several have at least a general (and some even a specific) focus on RFID. Table A-1 lists such RFID news sources. These include periodicals, Web sites, and associations dedicated to the task of disseminating the current trends, events, product information, and case studies of real-world RFID implementations, among other things. As mentioned previously, the list of these news sources is not exhaustive.

Table A-1   RFID News Sources

Name	Contact Information
Aim Global	www.aimglobal.org
Extreme RFID	www.extremerfid.com
Global ID Magazine	www.global-id-magazine.com
Inbound Logistics	www.inboundlogistics.com
Integrated Solutions	www.integratedsolutionsmag.com
RF Design	www.rfdesign.com
RFIDa	www.rfida.com
RFIDbuzz	www.rfidbuzz.com
RFID Gazette	www.rfidgazette.org
RFID Insights	www.rfidinsights.com
RFID Journal	www.rfidjournal.com
RFID News	www.rfidnews.org
RFID News & Solutions	www.rfidnas.com
RFID Talk	www.rfidtalk.com
Transponder News	www.transpondernews.com
Unwatched.org	www.unwatched.org/rubrique2.html
Using RFID	www.usingrfid.com

A.3   RFID Conferences

RFID conferences provide attendees with an excellent introduction to the world of technology adopters, vendors, and experts. Conference attendees can also often access practical, hands-on knowledge from implementers and users of the technology. Table A-2 lists RFID conferences. As mentioned previously, this list is not exhaustive.

Table A-2   RFID Conferences

Name	Contact Information
Canadian RFID Conference	www.rmoroz.com
EPCglobal Conferences	www.epcglobalinc.com
IDC Asia/Pacific RFID Conference	www.idc.com.sg
PMMI (Packaging Machinery Manufacturers Institute) RFID Conference	www.pmmi.org
RFID Commerce	www.scievents.com/rfidcom
RFID Devcon	www.shorecliffcommunications.com/rfidepcdev
RFID Europe Conferences	www.shorecliffcommunications.com/rfideu
RFID Journal LIVE	www.rfidjournallive.com
RFID ROI Summit	www.rfid.access-events.com
RFID Supply Chain Solutions	www.softmatch.com/conference.htm
RFID World	www.rfid-world.com
Smart Labels USA	www.smartlabelsusa.com
Wisconsin RFID Conference	www.uwebi.org/RFIDConference

Passive Tag Manufacturing Overview

The passive tag manufacturing process is currently a subject of intense activity as vendors strive to bring cheaper tags with more capabilities to the market. The traditional process of tag manufacturing is costly compared to the estimated cost for producing the widely anticipated 5¢ tag. The volume of such tags is anticipated in the billions per year. At a minimum, the output capacity of the existing manufacturing processes must increase in the order of a magnitude to meet this demand. This appendix examines the basic tag-manufacturing process. Most likely, the steps in this process will not change drastically even as the tag-manufacturing process changes, unless a radical process innovation occurs (for example, printing a complete tag on the item with a conductive ink).

The basic tag manufacturing steps are as follows:

1. Create the die.
2. Produce the tag antenna.
3. Create the inlet.
4. Convert the inlet.

The following sections discuss these steps.

B.1   Die

A *die* consists of an individual tag microchip on a silicon wafer. A silicon wafer is a circular slice of silicon a few inches in diameter (generally 8 inches) on which a significant number of microchips can be produced.

This manufacturing step generally consists of the following substeps:

1. Produce microchips on the silicon wafer.

    Chemical etching processes (also known as *photolithographic processes*) are used to produce several hundred to thousands of tag microchips on the silicon wafer.

2. Test.

    The individual microchips on the wafer are tested for functionality. Metal points are generally used in this testing to access the chip directly via its specialized contact fields. The chip is placed in a *test mode*, in which all its functionalities can be tested comprehensively. All defective chips are marked with a color code so that these can be identified and removed later. After the testing has been completed, the test mode of the microchip is permanently deactivated by blowing specific fuses on the chip (which prevents any unauthorized data access at a subsequent stage).

3. Embed unique serial numbers.

    This step is generally performed with the previous (testing) phase where a unique serial number is programmed into the chip's memory (for example, EEPROM). For a *read-only* (RO) tag, this is done at the hardware level by blowing a set of fuses of the microchip with a fine-pointed laser beam.

4. Separate individual microchips (dies).

    Typically, a diamond saw is used to cut the wafer up and separate the individual microchips on the wafer. However, these saws also destroy a part of the silicon that might have been used for producing more microchips. Therefore, alternative methods may also be used, especially when the microchips are very small (for example, 300 microns or less).

5. Embed a die into a substrate module.

    This is an optional step. Individual dies are embedded into a module on some kind of a substrate (for example, plastic). If the tags are too tiny, this step might not be feasible in terms of cost and capabilities of existing equipment. In this case, alternative methods need to be developed.

B.2   Tag Antenna

This step differs depending on the frequency of the passive tag produced. For *low-frequency* (LF) and *high-frequency* (HF) tags, very thin metal (for example, copper) wire is generally used to produce the antenna. Etched metal (for example, copper or aluminum) is generally used for manufacturing UHF and microwave tag antennas. In these cases, an antenna is also produced by printing it with a conductive ink that contains copper, carbon, or nickel. Different companies may design the tag antenna and the tag microchip.

B.3   Inlet

This manufacturing step consists of the following substeps:

1. Connect the module or the die (if the module was not produced) to the tag antenna on a substrate.

    This step is performed rapidly and automatically using special equipment (for example, robotic hands). The resulting assembly is called an *inlet* or an *inlay*.

2. Test.

    In this step, the inlet is tested in a contactless manner. Unsuccessful inlets are discarded.

    This stage produces a fully functional tag that is in a nonusable form.

B.4   Inlet Conversion

This step converts an inlet into a form that is usable and/or customized to specific ruggedness requirements. For example, the inlet could be sandwiched between two layers of paper or plastic to form a smart label. The top layer might contain areas for printing the sender and recipient addresses, the item bar code, a short textual description of the item, and so on. The bottom layer could consist of an adhesive backing that could be used to attach the label to an item. Alternatively, the inlet can be housed inside a glass capsule, plastic molding, or some other usable form. After this stage, another round of contactless testing is generally performed for quality assurance. The defective tags are discarded.

Tag converter companies can convert an inlet into a desired usable form.

At this point, the tag is ready to be used by a customer. Figure B-1 shows the typical tag-manufacturing process.

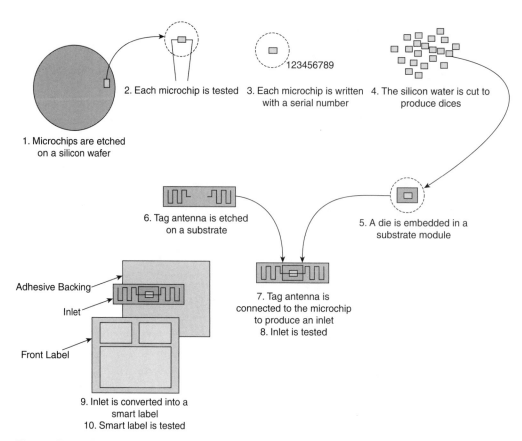

2. Each microchip is tested

3. Each microchip is written
   with a serial number

4. The silicon water is cut to
   produce dices

1. Microchips are etched
   on a silicon wafer

6. Tag antenna is etched
   on a substrate

5. A die is embedded in a
   substrate module

Adhesive Backing

Inlet

7. Tag antenna is
connected to the microchip
to produce an inlet
8. Inlet is tested

Front Label

9. Inlet is converted into a
smart label
10. Smart label is tested

Figure B-1   A typical tag manufacturing process.

Glossary

2G Second Generation. In mobile telephony, 2G protocols use digital encoding and support limited data communication capabilities. Digital technologies such as GSM, TDMA, and CDMA are examples of 2G technology.

2.5G Second and a half Generation. Extension of 2G wireless systems to increase data bandwidth and provide additional features such as packet-switched connection (for example, GPRS).

2-wire folded dipole A folded dipole antenna consisting of two straight conductors connected in parallel.

3G Third Generation. The next generation of wireless digital communication technology offering a substantially higher data rate than 2.5G.

3-wire folded dipole A folded dipole antenna consisting of three straight conductors connected in parallel.

active tag An RFID tag that has an on-board power supply (for example, a battery) and electronics to perform specialized tasks. An active tag uses its on-board power to transmit data; it does not use the emitted power of a reader. An active tag can therefore communicate even in the absence of a reader. A read range of 300 feet (91 meters approximately) or more can be achieved using an active tag. These tags are typically more expensive than passive tags.

actuator A mechanical device for control and movement of some object.

addressability The ability to address (that is, read and write) the individual memory fields of a tag's microchip.

agile reader An RFID reader that can operate in different frequencies or uses different tag-to-reader communication protocols.

AIAG Automotive Industry Action Group. A nonprofit association, AIAG's primary goals are to reduce cost and complexity within the automotive supply chain and to improve speed to market, product quality, employee health and safety, and the environment.

Air Interface Protocol A set of rules that governs the tag reader communication.

amplitude The height of a crest or the depth of a trough from the undisturbed position.

amplitude at a certain point Of a wave is the height or depth of this point from the undisturbed position, and is called positive or negative accordingly.

amplitude modulation Modulation of the wave amplitude.

annunciator An electrically controlled signaling or indicator device.

ANSI American National Standards Institute. It is a private, nonprofit organization that administers and coordinates the U.S. voluntary standardization and conformity assessment system.

antenna gain In simple terms, an increase in antenna gain means an increase in RF energy output of the antenna in a particular direction or pattern. Antenna gain is measured in decibels, abbreviated as dB.

anti-collision A mechanism that is used to mediate tag collisions so that a reader can read multiple tags in its read zone.

ASCII American Standard Code for Information Interchange. This is a standard-based representation of 256 characters that includes the uppercase and lowercase letters from the English alphabet, digits, punctuation marks, and control codes.

ASN Advanced Ship Notice. An EDI transaction that provides advance shipment data to the recipient. The recipient can then use this data to optimize the received logistics.

attenuate Decrease in the transmission strength of an antenna's signal.

attenuator A small device that, when attached to the reader antenna port, attenuates the antenna signal (usually by a fixed amount).

Auto-ID The mechanism by which a physical object can be identified in an automatic manner. Some examples of Auto-ID are bar code, RFID, biometric (for example, using fingerprint and retinal scan), voice identification, and optical character recognition (OCR) systems.

bar code A scheme by which textual information is presented as a printed symbol. A bar code can be either one dimensional or two dimensional.

battery-assisted tag A semi-passive tag.

carrier frequency Refers to the frequency at which the reader operates.

CDMA Code Division Multiple Access. This is a 2G radio communication technology that uses the spread spectrum scheme to efficiently utilize the frequency spectrum of transmission.

CEPT   Conférence Européenne des Postes et des Télécommunications (or The European Conference of Postal and Telecommunications Administrations). Established on June 26, 1959, CEPT is an association of European Telecommunications service providers.

checksum   A numeric value that is computed and stored for a block of data in tag memory. When a reader receives the data block, it computes a new checksum for the received data and checks it against the stored checksum value to ensure that this data block has been correctly transmitted.

choke point   A spot or location where a tag can be read while in transit.

circular polarized antenna   An antenna from which electromagnetic waves radiate in a circular pattern. This type of antenna has a shorter read range compared to a linear polarized antenna. In contrast to a linear antenna, however, a circular antenna is largely insensitive to tag orientation in its read field.

closed-loop system   An RFID system that is used inside the four walls of a business. Because the information about the tagged object is never shared outside the business, such a system can be proprietary (that is, might not use open standards).

commissioning a tag   When a tag is created and is uniquely associated with an object.

constructive interference   This occurs in case of a multipath when an original reader antenna wave is imposed on a reflected wave with an exactly matching waveform resulting in strengthening of the original signal.

contactless smart card   A special type of passive RFID tag that is read when in close proximity of a reader. No physical contact between the tag and the reader is necessary for reading.

coupling   A mechanism based on which the transfer of energy takes place from a reader to a tag.

coupling element of a reader   A reader antenna as it creates a electromagnetic field to couple with the tag.

CPGs   Consumer Packaged Goods. These classes of goods are consumed and need to be replenished periodically. Examples are food items, beverages, apparel, and cleaning goods.

CRC   Cyclic Redundancy Check. A type of checksum.

crest   The highest point of a wave.

CW   Continuous Wave. This is a radio wave with constant frequency and amplitude. From a communication's viewpoint, a CW does not have any embedded information in it, but can be modulated to transmit a signal.

cycle   One complete wavelength of oscillation of a wave.

data transfer rate   The maximum rate at which a reader can read the data from a tag. It is expressed in bits/second or bytes/second.

dB   Decibel. A unit for measuring antenna gain and other types of quantities such as reader power output.

decommissioning a tag   When a tag is disassociated with the tagged object.

destructive interference   This occurs in case of a multipath when an original reader antenna wave is imposed on a reflected wave with exactly the opposite waveform resulting in cancellation of the original signal.

die   Consists of an individual tag microchip on a silicon wafer. Plural: dice.

dipole   A type of tag antenna consisting of a straight electric conductor of a specific length that is interrupted at the center. The total length of a dipole antenna is half the wavelength of the frequency used.

discover   A reader is said to discover a tag when it reads a new tag that is not on its tag list.

discovery time   Time when a tag is discovered by a reader.

DoD   U.S. Department of Defense.

dollarization   The process by which a foreign country (for example, Ecuador) adopts the U.S. dollar as its national currency.

duty cycle   The time for which a reader can emit RF-energy to read tags.

EAN   European Article Numbering. A popular bar code symbology standard used in Europe. This is the European equivalent of UPC symbology.

EAN International   European Article Numbering International. A nonprofit European organization that administers EAN.

EAS   Electronic Article Surveillance. Technology for detecting item theft at retail stores, libraries, and so on. Simple RF devices that can either be turned on or off are generally used for this purpose.

ECC   Electronic Communications Committee. ECC is the committee that brings together the radio- and telecommunications regulatory authorities of the 45 CEPT member countries.

EDI   Electronic Data Interchange. An ANSI X12-based standard format used for exchanging business data such as ASN, invoices, and purchase orders.

EEPROM   Electrically Erasable, Programmable, Read-Only Memory. An EEPROM retains its data even when the power is turned off.

EIRP   Equivalent Isotropic Radiated Power. A measure of a reader's antenna power used in the U.S. It is generally expressed in watts. EIRP = 1.64 ERP.

electromagnetic waves   Created by electrons in motion and consist of oscillating electric and magnetic fields. These waves can pass through a number of different material types.

electronic pedigree   A secure electronic record that contains information about the movement of a particular product through the supply chain.

electronic signature   A unique electronic identification code of an item.

EPC   Electronic Product Code. A license-plate type identifier that can uniquely identify an item in a supply chain. This is currently owned by EPCglobal.

EPCglobal A joint venture between UCC and EAN International to commercialize EPC.

ERO European Radiocommunications Office. It supports ECC of the CEPT.

ERP Effective Radiated Power. A measure of a reader antenna power used in Europe. It is generally expressed in watts.

ERP Enterprise Resource Planning. This set of activities enables a business to manage its various aspects (for example, order, inventory, and customer service).

ESD Electrostatic Discharge. ESD is a sudden flow of electric current through a material that is an insulator under normal circumstances. If a large potential difference exists between the two points on the material, the atoms between these two points can become charged and conduct electric current.

ETSI European Telecommunications Standards Institute. ETSI is an independent, nonprofit organization whose mission is to produce telecommunications standards for today and for the future.

excite A term that indicates a passive tag microchip drawing power from a reader's signal to properly energize itself.

factory programmed An RO tag whose identification data was written at the time of manufacturing. Once written, the data cannot be changed because it is etched in the tag microchip.

false read Phantom read.

far field The area beyond one full wavelength of the RF wave emitted from a reader antenna.

FCC Federal Communications Commission. The FCC is an independent U.S. government agency, directly responsible to Congress. The FCC was established by the Communications Act of 1934 and is charged with regulating interstate and international communications by radio, television, wire, satellite, and cable. The FCC's jurisdiction covers the 50 states, the District of Columbia, and U.S. possessions.

field programmable A WORM or RW tag whose identification and other data can be written at the time when it is needed by the tag consumer. Once written, the tag data may or may not be rewritten depending on whether the tag is RW or WORM, respectively.

firmware Software stored in the ROM of a device.

Flash A type of constantly powered nonvolatile memory that can be erased and reprogrammed in units of memory (called blocks).

folded dipole A type of tag antenna consisting of two or more straight electric conductors connected in parallel and each half the wavelength (of the used frequency) long.

FRAM Ferroelectric RAM. A type of random access memory that combines the fast read and write access with the capability to retain data when power is turned off. It also has low power requirements. Contrary to its name, FRAM does not contain iron, but compounds such as lead zirconate titanate (PZT).

frequency The number of cycles in a second. The frequency of a wave is measured in hertz (abbreviated as Hz), named in honor of the German physicist Heinrich Rudolf Hertz. If the frequency of a wave is 1 Hz, it means that the wave is oscillating at the rate of one cycle per second. It is common to express frequency in KHz (or kilohertz = 1,000 Hz), MHz (or megahertz = 1,000,000 Hz), or GHz (or gigahertz = 1,000,000,000 Hz).

frequency band A range of frequencies on the electromagnetic spectrum.

frequency hopping An RF transmission scheme that minimizes interference among several devices.

GIAI Global Individual Asset Identifier. This is used to identify immovable asset as well as fixed inventory of a business.

GLN Global Location Number. This is used for representing location, trading partners, and legal entities.

GPS Global Positioning System. Used to compute the geographical location at any point on Earth within some limits of accuracy (generally between 33 feet (10 meters) to 328 feet (100 meters); accuracy within 3.3 feet (1 meter) can be achieved using military-approved specialized equipment). Owned by and operated by the DoD, GPS is available for general use around the world.

GPRS General Packet Radio Service. This is a 2.5G enhancement of GSM that uses packet-switched transmission technology. It has a theoretical maximum transfer rate of 171.2 Kbps although the actual speed usually ranges between 56 to 114 Kbps. It is typically used for continuous connection to the Internet from mobile phones and computers.

GRAI Global Returnable Asset Identifier. This is used for numbering returnable assets such as drums, gas cylinders, and so on.

GSM Global System for Mobile communications. This is the most widely used 2G digital mobile technology in the world today. GSM uses TDMA and is implemented in the 800, 900, 1800, and 1900 MHz frequencies.

GTAG Global TAG. This is aimed at facilitating global supply-chain operations in the 862 to 928 MHz (UHF) band. It provides a technical foundation with canonical data sets and applications guidelines.

GTIN Global Trade Identity Number. This is a globally unique EAN-UCC number for identifying products and services.

HA High Availability. A system or a system component that is continuously operational for a high percentage of its deployment lifetime.

HDMA Healthcare Distribution Management Association.

HF High Frequency. HF ranges from 3 MHz to 30 MHz, with 13.56 MHz being the typical frequency used for HF RFID systems.

IFPMA International Federation of Pharmaceutical Manufacturers Association.

in-phase Signals in a multipath that exactly match the wave pattern of the original signal.

inlay An inlet.

inlet A fully assembled RFID tag that is not in a usable form.

interrogator A reader.

ISM Industrial, Scientific, and Medical. The 2.4 GHz frequency band that is accepted world-wide for operations of industrial, scientific, and medical equipment.

ISO International Organization for Standardization. ISO is a network of the national standards institutes of 146 countries, on the basis of one member per country, with a Central Secretariat in Geneva, Switzerland, that coordinates the system. ISO is a nongovernmental organization.

Kbps Kilobits (for example, 1,024 bits) per second.

LF Low Frequency. Frequencies between 30 KHz and 300 KHz are considered low, with the 125 KHz to134 KHz frequency range being commonly used by RFID systems. A typical LF RFID system operates at 125 KHz.

line of sight Unobstructed path.

linear polarized antenna RF waves emanate in a linear patter from this type of antenna. A linear polarized antenna is sensitive to tag orientation in its read field.

micron One thousandth (1/1000) of a millimeter.

microstrip antenna A patch antenna.

microwave frequency This ranges upward from 1 GHz. A typical microwave RFID system operates either at 2.45 GHz or 5.8 GHz, although the former is more common.

modulated backscatter The type of communication mechanism between a tag and a reader. In this case, a part of the RF waves emitted by the reader is modulated and reflected back by the tag.

modulation The process of changing the characteristics of a radio wave to encode some information-bearing signal. Modulation can also mean the result of applying the modulation process to a radio wave.

multipath RF waves from a reader antenna can get reflected in the presence of RF-opaque objects in the environment. These reflected waves are scattered and might arrive at the reader antenna at a different time via different paths. Waves that arrive in-phase with the original antenna signal enhance the signal, whereas those that arrive out-of-phase cancel it.

near field The area between a reader antenna and one full wavelength of the RF wave emitted by the antenna.

negative amplitude The depth of a trough from the undisturbed position.

out-of-phase Signals in a multipath that are exactly the opposite of the wave pattern of the original signal.

passive tag An RFID tag that does not have an on-board power supply. A passive tag uses a part of the power emitted by a reader to transmit its data. Therefore, a passive tag can communicate only in the presence of a reader.

patch antenna An antenna consisting of a rectangular metal foil or plate mounted on a substrate such as Teflon.

period of oscillation The time taken by a wave to complete one cycle.

persist time The time for which a reader keeps information from a read tag on its tag list.

phantom read A spurious read of a nonexistent tag by a reader.

photolithographic process A chemical etching process used to produce tag microchips on a silicon wafer.

planar antenna A patch antenna.

POS Point of Sale. The physical location where a good is sold to a buyer.

positive amplitude The height of a crest from the undisturbed position.

power The strength of the radiated RF energy of a reader antenna.

read redundancy The number times a tag can be read when in the read zone of a reader.

read window Read zone.

read zone The three-dimensional region protruding from a reader antenna inside which a tag can be read.

reader A device that can read from and write data to compatible tags.

reader antenna Broadcasts the reader transmitter's RF signal into its surroundings and receives tag responses on the reader's behalf.

reader collision Occurs when the read zones of two or more readers overlap, resulting in interference of signal from one reader with that from another.

reader range Distance at which a tag can successfully communicate with a reader antenna.

reprogrammable The ability to rewrite a tag data. Also, an RW tag.

RF Radio Frequency.

RF-absorbent A material that allows RF waves to pass through it at the expense of substantial RF energy loss. This property is frequency dependent. Water is RF-absorbent at UHF and microwave frequencies.

RF-friendly An RF-lucent material.

RF-lucent A material that lets RF waves pass through it without any substantial loss of RF energy. This property is frequency dependent. Paper is RF-lucent at UHF and microwave frequencies. Metal is RF-lucent at LF and HF frequencies.

RF-opaque A material that blocks, reflects, and scatters RF waves. This property is frequency dependent. Metal is RF-opaque at UHF and microwave frequencies.

RF waves These are electromagnetic waves with wavelengths between 0.04 inches (0.1 centimeters) and 621.4 miles (1,000 kilometers). Another equivalent definition in terms of frequency is radio waves are electromagnetic waves whose frequencies lie between 30 Hz and 300 GHz. Other electromagnetic wave types are infrared, visible light wave, ultraviolet, gamma-ray, x-ray, and cosmic-ray. RFID uses radio waves that are generally between the frequencies of 30 KHz and 5.8 GHz.

RFID Radio Frequency Identification. A technology that uses radio waves to automatically identify physical objects.

RO Read-only. A tag that can be written only once in its lifetime. The data is burned into the tag at the factory during the manufacturing stage.

ROM Read-only Memory. A type of memory, which once written, can never be modified.

RW Read-write. A tag whose data can be rewritten multiple times.

SAW Surface Acoustic Wave. A technology that uses a low-power 2.45 GHz frequency to generate surface waves on a piezoelectric material. A part of this wave is converted back to an RF signal, which can then be received and decoded by a reader to extract the embedded data.

scanner A device that can transmit and receive RF waves. A scanner is an integral part of a reader.

semi-active tag A semi-passive tag.

semi-passive tag A semi-passive tag has an on-board power supply. This is used to power its on-board electronics, but is not used for data transmission. A semi-passive tag uses a part of the power emitted by a reader to transmit its data. Therefore, a semi-passive tag can communicate only in the presence of a reader.

sensor A device that can detect a physical stimulus. A large variety of sensors exist today (for example, temperature, mechanical, light, and chemical). A sensor can convert the detected signal into some kind of electrical or electronic signal.

signal attenuation Weakening of RF energy (for example, when it passes through an RF-absorbent material).

silicon wafer A circular slice of silicon a few inches in diameter on which a large number of microchips can be produced.

singulation A mechanism by which a reader can uniquely identify a single tag among several tags present in the read zone.

smart label This is a bar code label that has an RFID tag embedded in it.

SNMP Simple Network Management Protocol. A standard used for monitoring and management of network devices.

spread spectrum A form of wireless communications in which the frequency of the transmitted signal is intentionally varied. This results in a much greater bandwidth than the signal would have if its frequency were not varied.

SRD   Short-Range Devices. This device types include RFID.

SSCC   Serial Shipping Container Code. This is used to identify shipping units such as a pallet, case, carton, and so on.

sweethearting   When checkout personnel removes an item tag without accepting any payment or accepts a sum that is less than the item's sale price.

symbology   A method that translates textual information into a printed symbol for a bar code.

tag   A device that can generate, store, and transmit data to a reader in a contactless manner using radio waves.

tag alignment   Positioning of a tag antenna with respect to the reader antenna.

tag antenna   This is used for drawing energy from the reader's signal to energize the tag and for sending and receiving data from the reader. A tag antenna is physically attached to the tag microchip. The antenna geometry is central to the tag's operations.

tag collision   Occurs when more than one tag in the read zone of a reader attempts to communicate with the reader at the same time.

tag list   The list of tags last read by a reader.

tag orientation   Tag alignment.

TCO   Total Cost of Ownership. The sum total cost of owning a product by adding its purchase price and maintenance costs for the duration of ownership.

TDMA   Time Division Multiple Access. A digital wireless transmission scheme which assigns unique time slots to users within the channel to share access to a single radio frequency channel without interference. In context of RFID, this technique can be used for resolving reader collision. Using this technique, the readers read tags at different times, and thus can avoid any interference with one another.

transponder   A tag that can act both as a transmitter and a receiver. This tag only transmits when interrogated by a reader. The tag enters into a sleep or low-power state when not being interrogated by a reader. Thus, not all tags can be called transponders.

trough   The lowest point of a wave.

UCC   Uniform Code Council. A nonprofit organization that administers UPC in North America.

UHF   Ultra High Frequency. Frequencies between 300 MHz and 1 GHz are called ultra high frequencies. A typical passive UHF RFID system operates at 915 MHz in the U.S. and at 868 MHz in Europe. A typical active UHF RFID system operates at 315 MHz and 433 MHz.

UID   Unique Identification. This is a U.S. Department of Defense numbering scheme for asset tracking.

UPC   Uniform Product Code. A popular bar code symbology standard used in North America.

wave   A disturbance that transports energy from one point to another.

wavelength   The distance between two consecutive crests or two consecutive troughs.

WHO World Health Organization.

WMS Warehouse Management System. A system used for managing warehouse processes and operations.

WORM Write Once, Read Many. A passive/active tag that can be written only once in its lifetime. This tag can only be read after it has been written.

write rate The rate at which data is written to a tag. This involves data transfer from the reader to the tag, writing to the tag memory, and validating that the data has been correctly written.

Index

Wouldn't it be great

if the world's leading technical publishers joined forces to deliver their best tech books in a common digital reference platform?

They have. Introducing
InformIT Online Books
powered by Safari.

Specific answers to specific questions.
InformIT Online Books' powerful search engine gives you relevance-ranked results in a matter of seconds.

Immediate results.
With InformIT Online Books, you can select the book you want and view the chapter or section you need immediately.

Cut, paste and annotate.
Paste code to save time and eliminate typographical errors. Make notes on the material you find useful and choose whether or not to share them with your work group.

Customized for your enterprise.
Customize a library for you, your department or your entire organization. You only pay for what you need.

Get your first 14 days FREE!
For a limited time, InformIT Online Books is offering its members a 10 book subscription risk-free for 14 days. Visit **http://www.informit.com/online-books** for details.

Safari
POWERED BY
TECH BOOKS ONLINE

InformIT Online Books

informit.com/onlinebooks

inform IT

www.informit.com

YOUR GUIDE TO IT REFERENCE

Articles

Keep your edge with thousands of free articles, in-depth features, interviews, and IT reference recommendations – all written by experts you know and trust.

Online Books

Answers in an instant from **InformIT Online Book's** 600+ fully searchable on line books. For a limited time, you can get your first 14 days **free**.

Safari
TECH BOOKS ONLINE

Catalog

Review online sample chapters, author biographies and customer rankings and choose exactly the right book from a selection of over 5,000 titles.